INTRODUCTORY STATISTICS ON THE TI-83

WITH NOTES AND PROGRAMS FOR THE TI-82

JOHN G. HORNER
Cabrillo College

VIRGINIA DEUS
Diablo Valley College

John Wiley & Sons, Inc.
New York • Chichester • Weinheim • Brisbane • Singapore • Toronto

COVER PHOTO Mark Tomalty/Masterfile

Permission has been granted to John Wiley & Sons, Inc. by Texas Instruments Incorporated to use any screenshots and any excerpts from the Texas Instruments manuals that may appear in this work. The TI-82 and TI-83 calculators are trademarks of Texas Instruments Incorporated.

ISBN 0-471-19855-2

Printed in the United States of America

10 9 8 7 6 5 4 3 2

Printed and bound by Hamilton Printing Company

Preface

Learning elementary statistics and simultaneously learning how to do computer calculations is an intimidating combination for most students. The **TI-83** and this book can be real "stress relievers." This manual has been written to show you *how* to use, step by step, the statistical functions on the **TI-83**. A good beginning statistics book, such as Mann's Introductory Statistics, 3rd ed., John Wiley & Sons, Inc., can explain the *why* of these functions. This manual has references to the Mann text at the bottom of most pages to assist you as you study. To overcome confusion and be successful in learning statistics, practice the examples in this book, do your assignments on a punctual basis, and ask your classmates and your instructor for help. We have made every effort to make the examples in this manual as clear as possible, but only your practice can make them clear to you.

If you do not yet own a calculator, buy a **TI-83**. It has functions available on no other Texas Instruments' machine and was especially designed to help with statistics. However, if you already own a **TI-82**, programs in this manual can enable your machine to do almost everything you will need. See page 99 for details. This manual was written to help you with either of these machines. In most cases, instructions are identical. On examples where the calculators differ, separate instructions are given.

The subjects in this manual are taught by example. Each example includes a step-by-step table. The first column details the exact key strokes you need, while the second gives brief explanations, and the third column shows pertinent screens.

To Instructors

This manual is the result of student requests for handouts. Over several semesters we have developed a format from which independent learning is easy. Therefore, the instructor can focus lecture time on the important ideas of statistics with minimum time spent on how-to button pushing. With this manual and the Texas Instruments calculators, you can easily incorporate some very sophisticated problems into your course. For example, the **TI-83** can support lists with 999 entries. For many of your assignments, you will find the calculator more convenient and much easier to manage than a conventional computer lab/software environment.

The **TI-83** is the more powerful machine, and any student acquiring a new calculator should choose it. The **TI-82** and **TI-83** are virtually the same for descriptive statistics and linear regression. We strongly recommend using the provided programs for the binomial, normal, and Poisson distributions as well as for the inverse normal function. These programs have interfaces that are the same for both machines. Confidence intervals and hypothesis tests are handled differently. The **TI-83** uses built in functions while the **TI-82** requires the programs described on page 99.

We have tried to make programs available without your having to type them. Please visit our Web Sites for further information. See page 81 for details.

Acknowledgments

We wish to thank our editor, Barbara Bredenko, without whose guidance and help we could not have finished this manual. Her expertise and encouragement were always welcome.

We wish to thank our many colleagues and students who tested our materials and reported what was really helpful. Thoughtful suggestions and comments that also shaped the manual were made by the following reviewers:

Greg Perkins	Peter J. Nodzenski	James Daniels
Hartnell College	Black Hawk College	Palomar Community College

We owe a special thank you to Barbara Kliejunas, Jeff Mock, Michael Lloyd, and Kevin Deus for sharing programming ideas and supplying computer support.

We owe an unexpected thank you to Newton, a Dalmatian, who insisted that we leave the computers from time to time to take a walk.

Finally, we want to give a special heart-felt thank you to Mr. Gordon Wall who years ago taught us the beauty of mathematics and introduced us to statistics.

John G. Horner
Cabrillo College
Aptos, California

Ginny Deus
Diablo Valley College
Pleasant Hill, California

Table of Contents

Getting Started

Descriptive Statistics

Probability Distributions

One Population: Estimation and Hypothesis Testing

One Population: Estimation and Hypothesis Testing

Two Populations: Estimation and Hypothesis Testing

Chi-Square Tests

ANOVA

Linear Regression

Programs

Appendix

TI-83
TEXAS INSTRUMENTS
Y1=4sin(X)
Y=3.9658
X=1.7017
STAT PLOT
Y=
TBLSET
WINDOW
FORMAT
ZOOM
CALC
TRACE
TABLE
GRAPH
QUIT
MODE
INS
DEL
2nd
LINK
LIST
STAT
A-LOCK
ALPHA
DISTR
VARS
CLEAR
DRAW
PRGM
ANGLE
MATRX
TEST
MATH
TAN-1
TAN
COS-1
COS
SIN-1
SIN
FINANCE
LOG
LN
RCL
STO
ENTRY SOLVE
ENTER
ANS
(-)
MEM
CATALOG
OFF
ON

Basics

Turning the Calculator ON and OFF

The [ON] button is in the lower left corner of the calculator. To turn the calculator **OFF**, you need to press two buttons. First press the [2nd] key in the upper left corner, then press the [ON] button. Notice the word **OFF** above the [ON] button. To use any of the yellow functions [blue on the **TI-82**], printed above the keys, press the [2nd] key first, then the key just below the function you want. To use any of the green functions or letters [white on the **TI-82**], printed above the keys, first press the [ALPHA] key.

Darkening and Lightening the Screen

You will need to adjust your screen brightness depending on lighting and the condition of your batteries. To darken the screen, first press the [2nd] key and then the ▲ key. You may need to repeat this key combination several times. To lighten the screen, press the [2nd] key and the ▼ key alternately and repeatedly, until the screen lightens. If you still have trouble adjusting the brightness, your batteries may need replacing. After you have replaced weak batteries, adjust the screen brightness as necessary.

Stopping the Machine

When you want to stop using a function or menu, you need to **QUIT**. Press [2nd], then the [MODE] key. This combination gives you the **QUIT** function.

Editing

Press the [CLEAR] key to clear one line of instructions. If the cursor is on a blank line on the home screen, press the [CLEAR] key to clear the entire screen. Using the arrow keys to first highlight an entry, press the [DEL] key to remove only that one entry. To insert within a previously written line, use the arrow keys and then the [2nd], **INS** combination. To recall your last command for editing or re-execution, use the [2nd], [ENTER] combination.

Screen Display

If your screen looks "**w*e*i*r*d**," check settings under [MODE] and also under [2nd] **FORMAT** [**WINDOW>FORMAT** on **TI-82**]. The settings usually used are listed at the left side of the screen and should be highlighted. Use the arrow keys to highlight the desired settings, and the [ENTER] key to make changes. (See page 84 for more details.)

Entering Data

Example: Enter the following data into List 1.

12, 13, 15, 12, 17, 20, 21, 25, 22, 27

FIRST: Clear any data remaining from previous problems.

Steps

STAT

1 Select **EDIT**.

L1 Move cursor to highlight **L1**. (Screens will vary.)

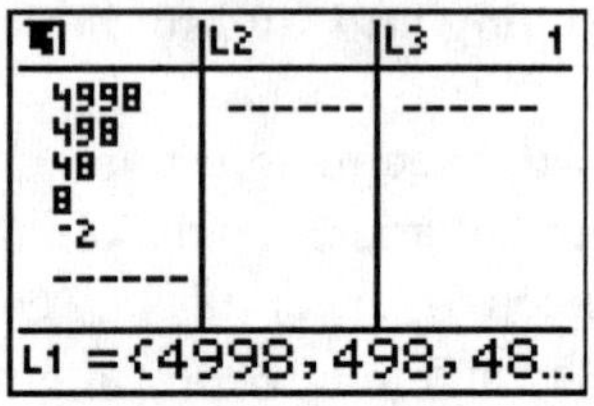

CLEAR

ENTER Editor is now ready to input List 1.

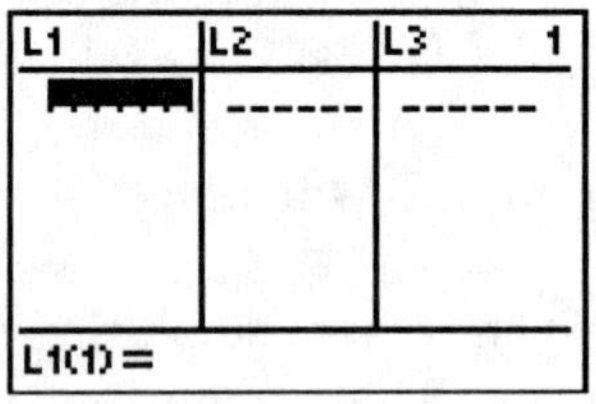

SECOND: Enter the data.

Steps

Enter the values Enter the values given above. Press **ENTER** after each entry. Notice that the calculator is awaiting the 11th entry, meaning you have input 10 numbers.

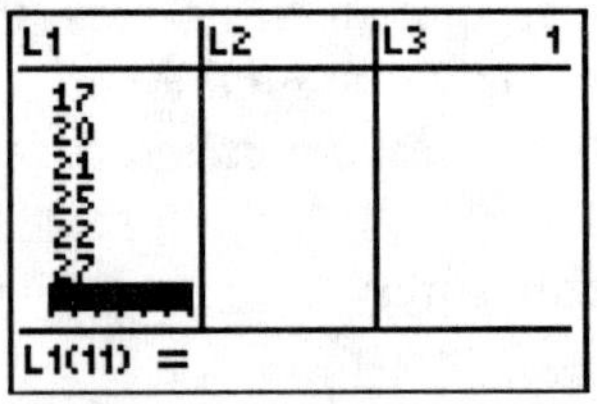

2nd QUIT

Editing Data

Example: In the previous example, the following data were entered into List 1.

12, 13, 15, 12, 17, 20, 21, 25, 22, 27

Into this list insert 11 at the beginning, delete the 17 and 20, and change the 22 to 32.

Steps

STAT	
1	Select **EDIT**.
Move cursor to highlight 12	

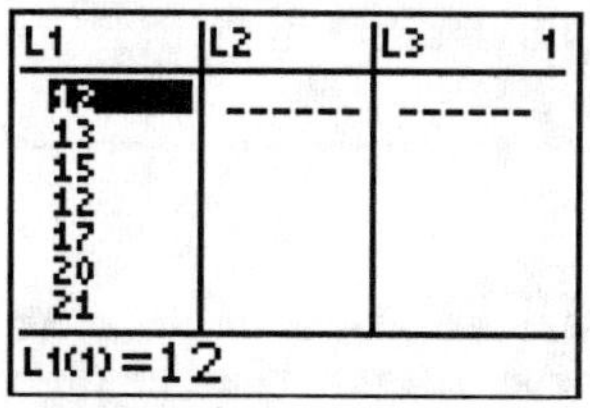

2nd INS	Insert a **0** into the first entry of List 1 so all other entries are "pushed down."

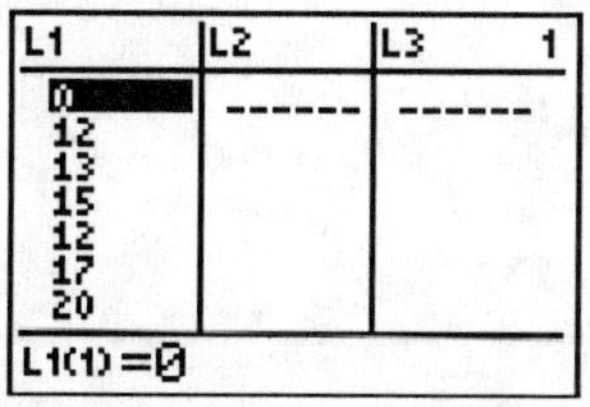

11 ENTER	Overwrite **0** with **11**.
▼▼▼▼	Move cursor down to highlight **17**.
DEL	Delete **17** so that the following entries "move up."
DEL	Delete **20**.
▼▼	Move cursor down to highlight **22**.
32 ENTER	

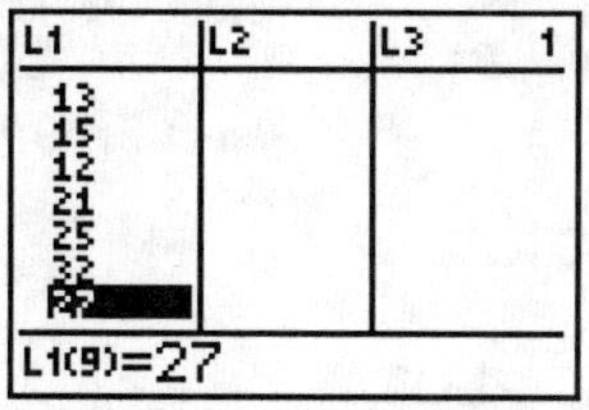

▼	Notice that after adding 1 entry, deleting two entries, and changing 1, there are now 9 entries in List 1.

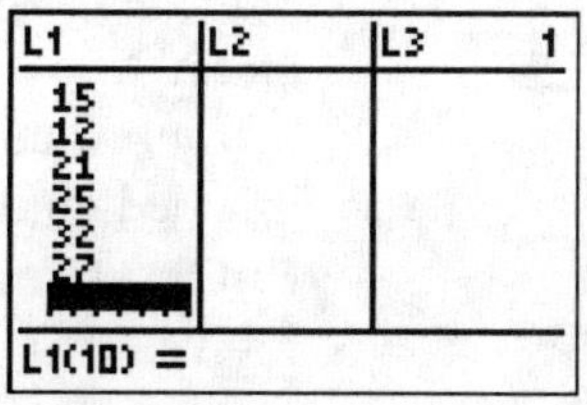

Histograms

Example: The following data represent quiz scores for a math class.

78 84 90 85 94 96 38 97 93 88 96 59 76 92 93 79 85 96 67 80 88
92 93 95 89 98 93 84 83

Construct a ***histogram*** for these data using 29.5, 39.5, 49.5, 59.5, 69.5, 79.5, 89.5 and 99.5 as cutoff points.

FIRST: Clear **L1**, then input the data into **L1**. Double check all 29 entries.

SECOND: Set the window to use the cutoff points shown below.

Steps

WINDOW		
Enter settings	Use the cursor keys to move to a setting, type over the setting and press **ENTER**. Your screen should look as shown.	WINDOW Xmin=29.5 Xmax=99.5 Xscl=10 Ymin=-3 Ymax=15 Yscl=1 Xres=1

THIRD: Turn off the graphs of any functions or **STAT PLOT**s currently turned **ON**.

Steps

VARS	[On **TI-82**, use **2nd Y-VARS**.]	
▶	Select **Y-VARS**.	
4 [**5 on TI-82**]	Select **On/Off….**	
2	Select **FnOff**.	
ENTER		
2ND		
STAT PLOT		
4	Select **PlotsOff**.	
ENTER	**Note:** Any old functions will be retained in your calculator, but they will not be graphed until you issue the command **FnOn**. Use the **PlotsOn** command to revive plots.	FnOff Done PlotsOff Done

FOURTH: Construct the histogram on the screen and **TRACE** the result.

Steps		
2nd **STAT PLOT**		
1	Select the first plot.	
Enter **Settings**	Use the cursor keys to move to a setting, and press **ENTER** to select it. Your screen should look as shown. The **Xlist:** setting and the **Freq:** setting must be typed if a change is needed. [**TI-82** screen differs slightly.]	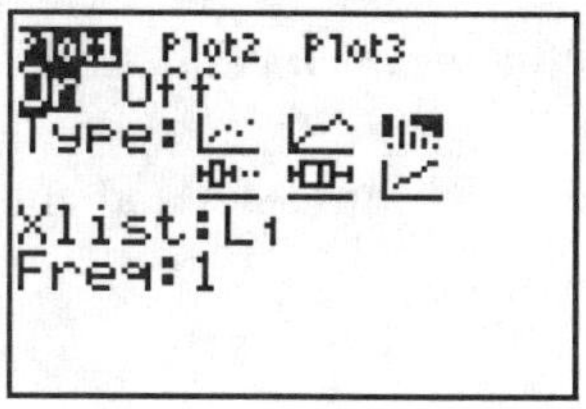
GRAPH	This is the Histogram using the data from List 1.	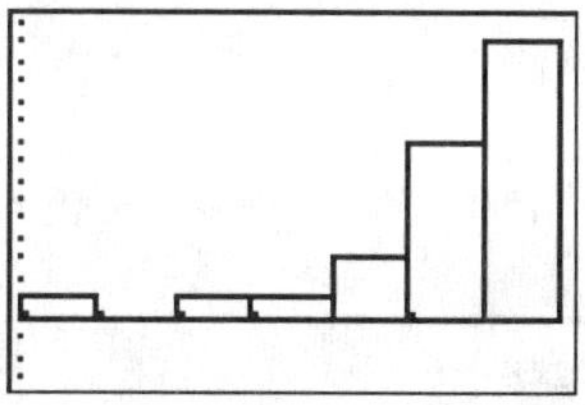
TRACE	Activating the **TRACE** key will show the minimum and maximum of the interval, and the frequency. In this example, there is one person ("**n=1**") with a quiz score between "**min=29.5**" and "**max<39.5**."	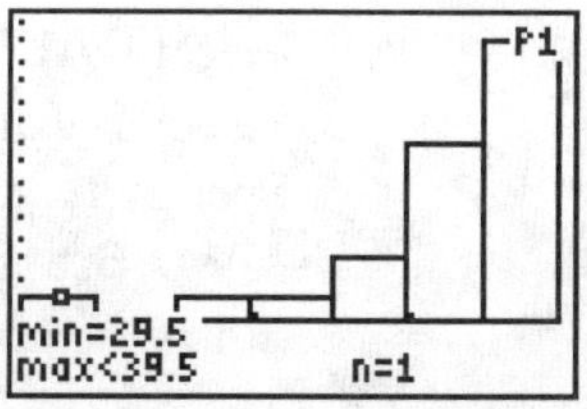
▶▶▶▶▶	As an example, moving the blinking cursor five classes to the right shows there are nine persons ("**n=9**") with quiz scores between "**min=79.5**" and "**max<89.5**."	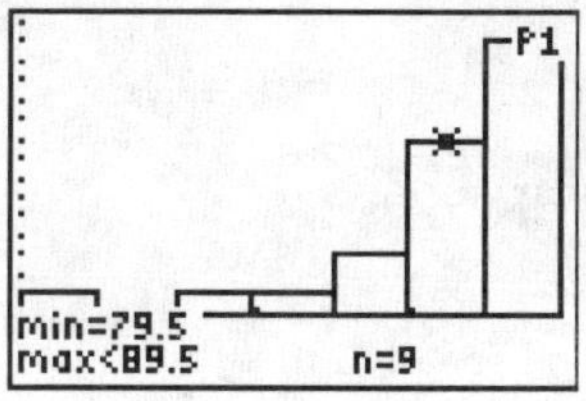
CLEAR **CLEAR**	End tracing. Turn off the histogram.	

Mean, Median, and Standard Deviation from Raw Data

Example: The following numbers represent scores on a statistics test for a population of 12 students.

88 75 70 98 95 66 42 77 71 79 68 50

Find the ***mean***, ***median*** and ***standard deviation*** of these data.

FIRST: Clear List 1 and enter data.

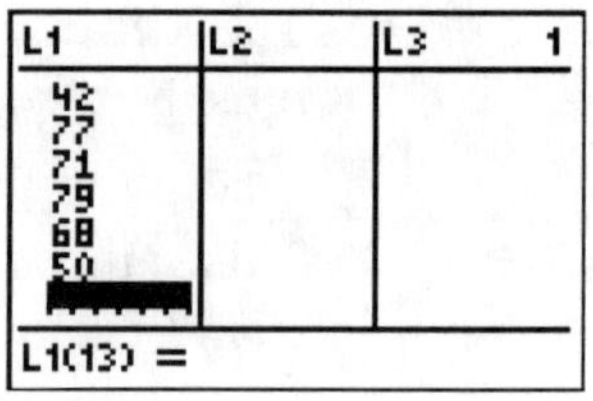

SECOND: Calculate the summary statistics.

Steps

STAT		
▶	Select **CALC**.	
1	Select **1-Var Stats**.	
2nd	Specify for which List the	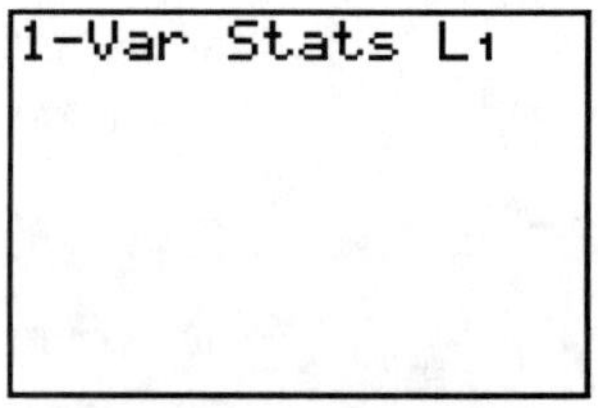
L1	**1-Var Stats** are to be calculated. (**L1** is located above the **1** key.)	
ENTER		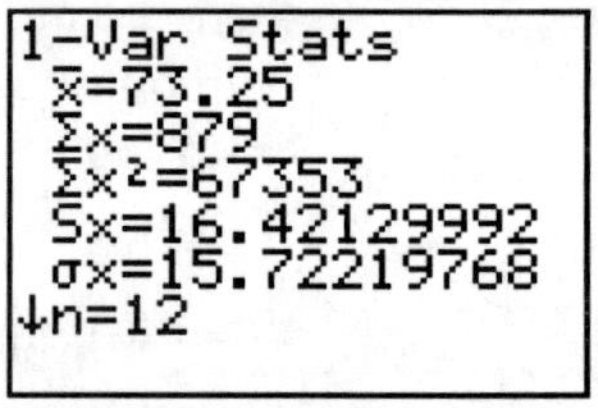
▼▼▼▼▼	Scroll down and find the median.	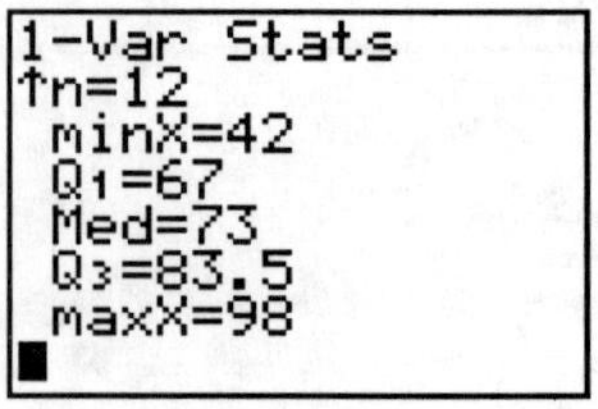

Solution: $\mu = 73.25$, $\sigma = 15.722$, median $= 73$

Note: variance $= \sigma^2 = 247.1875$.

Mann: 3.1, 3.2

Mean, Median, and Standard Deviation from Data With Frequencies

Example: The following table gives quiz scores and the number of students with each score for a total of 30 students. Find the mean, median, and the standard deviation for this sample.

[The **TI-82** calculator cannot do frequencies greater than 99.]

Score on quiz	No. of students (frequency)
10	6
9	4
8	3
7	8
6	2
5	1
4	1
3	2
2	1
1	2

FIRST: Input the quiz scores into List 1 (**L1**), and the related frequencies into List 2 (**L2**), after deleting any old data.

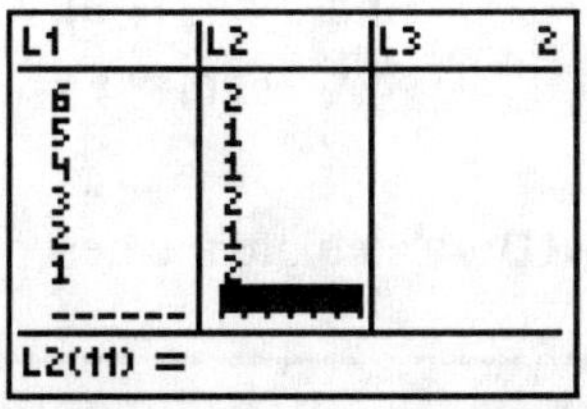

SECOND: Calculate the descriptive statistics using **1-Var Stats**.

Steps

STAT	
▶	Select **CALC**.
1	Select **1-Var Stats**.
2nd	
L1	The **1-Var Stats** function requires values of **x** (**L1**), and then the associated frequencies (**L2**).
,	
2nd	
L2	
ENTER	
▼▼▼▼▼	View all results.

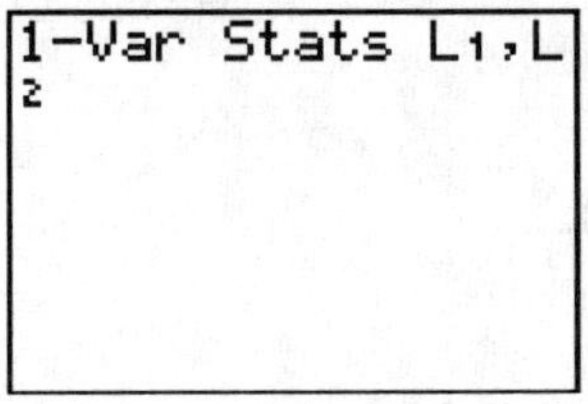

Solution: $\bar{x}$ = 6.9, s_x = 2.734, median = 7

Note: variance = s_x^2 = 7.4724

Mean and Standard Deviation from Grouped Data

Example. The following table shows the number of units taken this semester for a sample of 40 statistics students. Find the mean and standard deviation.

[The **TI-82** calculator cannot do frequencies greater than 99.]

No. of units	Frequency
1- 5	1
6-10	5
11-15	18
16-20	12
21-25	4

FIRST: Clear **L1**, and **L2**.

SECOND: Input the midpoints of each class into List 1 (**L1**), and the related frequencies into List 2 (**L2**). The calculator can help. For example, type the first entry in **L1** as **(1+5)/2** then **ENTER**. **3** will appear.

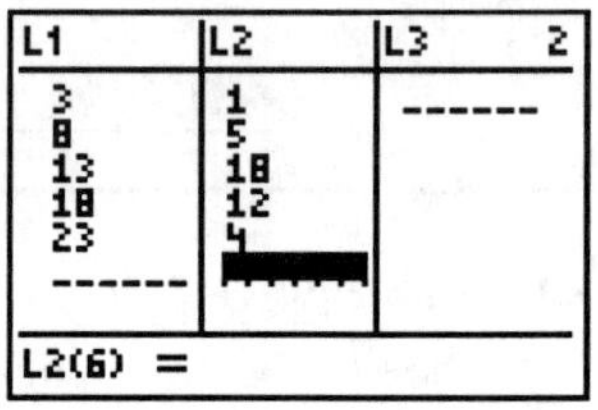

THIRD: Calculate the descriptive statistics.

Steps

STAT		
▶	Select **CALC**.	
1	Select **1-Var Stats**.	
2nd		
L1	The **1-Var Stats** function requires values of **x** (**L1**), and the associated frequencies (**L2**).	
,		
2nd		
L2		
ENTER		
▼▼▼▼▼	Scroll down to view all results.	

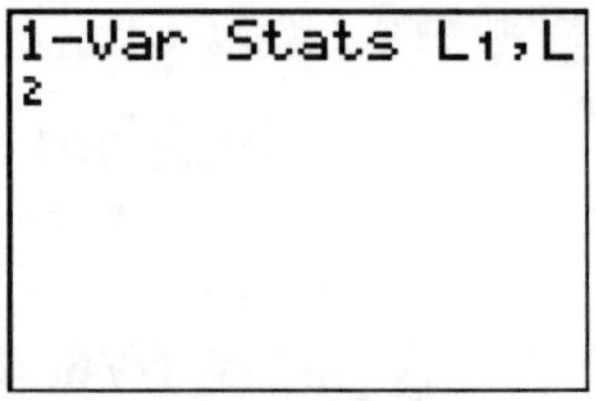

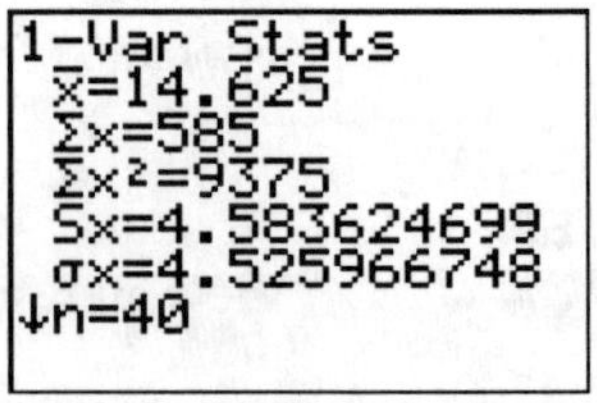

Solution: $\bar{x}$ = 14.625 $\quad$ s_x = 4.584 $\quad$ **Note:** variance = s_x^2 = 21.0096

Mann: 3.3

Sorting and the Mode

Although the calculator does not give the ***mode*** directly, it can sort the data which then makes it easier to find the mode.

Example: Find the mode of the following data set:

3	4	6	7	8	9	7	6	5	8	3	4
1	2	8	9	3	2	5	6	8	6	9	2

FIRST: Input the data into **L1**. Double check all 24 entries.

SECOND: Sort the data within **L1**.

Steps

STAT	
2	Select **SortA(** to choose an *ascending*-order sort.
2nd	Select the List to be sorted.
L1	
)	
ENTER	**L1** is now sorted.
STAT	View List 1.
1	

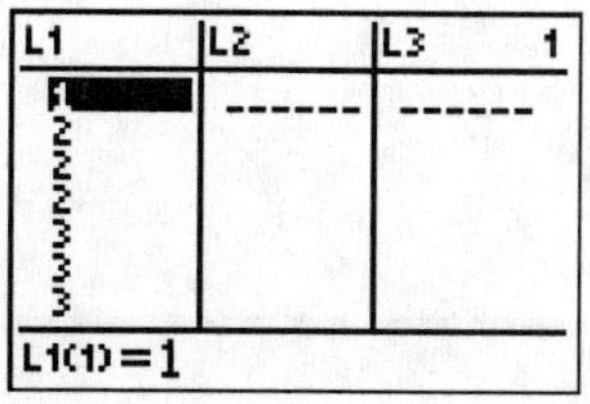

Now that the data list is organized, you can count frequencies. Scroll down through List 1, and count the frequency of each outcome. The frequency for each outcome is shown at the right.

Outcome	Frequency
1	1
2	3
3	3
4	2
5	2
6	**4**
7	2
8	**4**
9	3

Solution: The modes are 6 and 8.

Boxplots

Example: The following data represent quiz scores for a math class.

78 84 90 85 94 96 38 97 93 88 96 59 76 92 93 79 85 96 67 80 88
92 93 95 89 98 93 84 83

Construct a ***boxplot*** for these data.

FIRST: Clear any old data from **L1**, then input these scores into **L1**. Double check all 29 entries.

SECOND: Set the window to use the values shown below. (For other problems, note that **Xmin** should be less than the least value in the data set and **Xmax** should be greater than greatest value in the data set. Use the command **1-Var Stats** to get information on the range of **x**. The values of **Ymin** and **Ymax** are not used for boxplots.)

Steps

WINDOW		
Enter settings	Use the cursor keys to move to a setting, type over the setting and press **ENTER**.	WINDOW Xmin=29.5 Xmax=99.5 Xscl=10 Ymin=-3 Ymax=15 Yscl=1 Xres=1

THIRD: Turn off the graphs of any functions or **STAT PLOT**s currently turned **ON**.

Steps

VARS	[On **TI-82**, use **2nd Y-VARS**.]
▸	Select **Y-VARS**.
4 [**5** on **TI-82**]	Select **On/Off...**.
2	Select **FnOff**.
ENTER	
2ND	
STAT PLOT	
4	Select **PlotsOff**.
ENTER	

FOURTH: Construct the boxplot on the screen and **TRACE** the result.

Steps		
2nd **STAT PLOT**		
1	Select the first plot.	
Enter **Settings**	Use the cursor keys to move to a setting, and press **ENTER** to select it. Your screen should look as shown. The **Xlist:** setting and the **Freq:** setting must be typed if a change is needed. [**TI-82** screen differs slightly.]	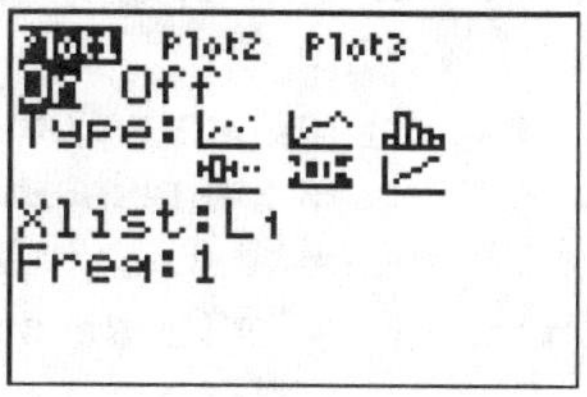
GRAPH	This is the boxplot using the data from List 1.	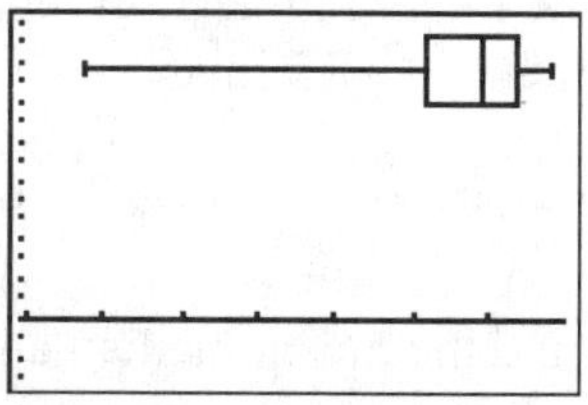
TRACE	Activating the **TRACE** key will show the minimum, Q_1, median, Q_3, and maximum of the data. In this example, the median of the data in List 1 is 89.	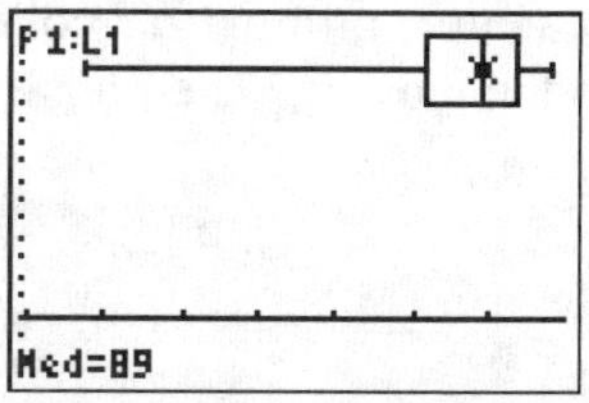
▶	Moving the blinking cursor once to the right shows that Q_3 (third quartile) = 93.5.	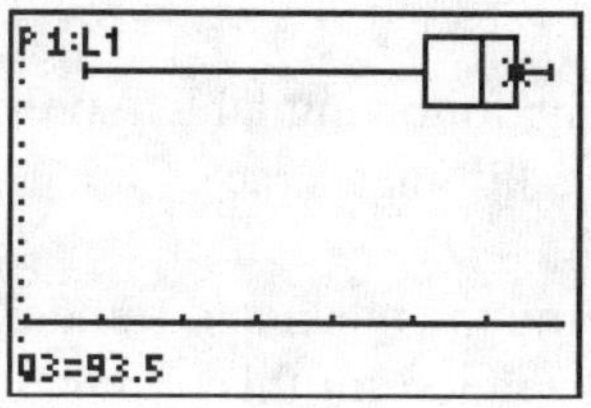
CLEAR	End tracing.	
CLEAR	Turn off the Boxplot.	

Supplementary Problems

1. Find the mean, median, and standard deviation for the following data set which represents the phone bills for a sample of 15 college students:

   $25.34 12.36 19.75 63.02 88.25 24.78 36.58 44.21
   16.78 12.54 17.54 18.63 25.12 37.95 12.34

2. Create a histogram for the data in problem 1 above, using 10.5, 30.5, 50.5, 70.5, and 90.5 as cutoff points.

3. Create a boxplot for the data in problem 1.

4. The following data set represents annual rainfall (in inches) for 10 randomly selected cities in California.

   12.65 10.21 6.23 25.67 32.98 16.58 17.58 23.66 25.44 13.54

   Find the mean, median and standard deviation for this data set.

5. A sample group of 27 college students were asked how many times a year they went to the movies. Their answers were as follows:

   5, 12, 25, 36, 55, 0, 6, 8, 5, 0, 1, 5, 2, 24,
   3, 16, 30, 12, 50, 0, 1, 5, 6, 3, 6, 10, 25

   Find the mean, median and mode for this data set.

6. The following table represents the times taken by all the runners who finished a 5 kilometer race.

   Find the mean and standard deviation for this population.

Time (min.)	**Number of runners**
.5-12.5	5
12.5-24.5	68
24.5-36.5	99
36.5-48.5	31
48.5-60.5	7

Solutions for Supplementary Problems

1.

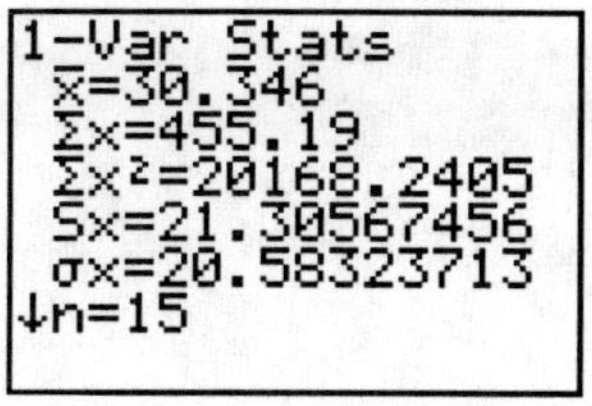

The mean is $30.35, the standard deviation is $21.31. (Scroll down to see median.)

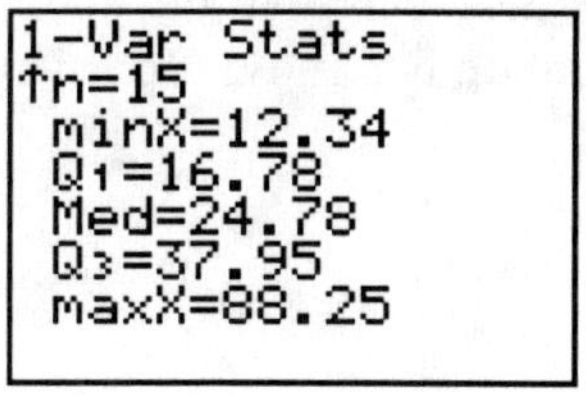

The median is $24.78.

2.

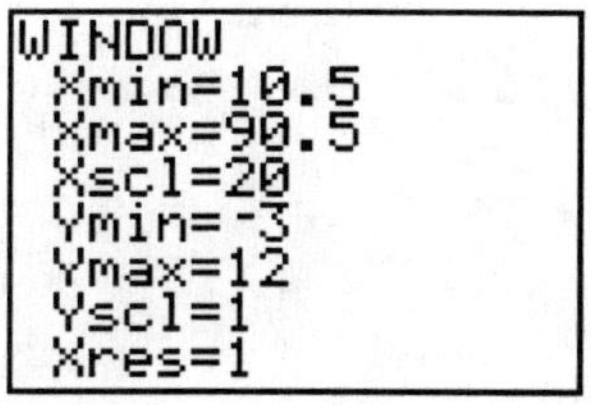

These are good window settings.

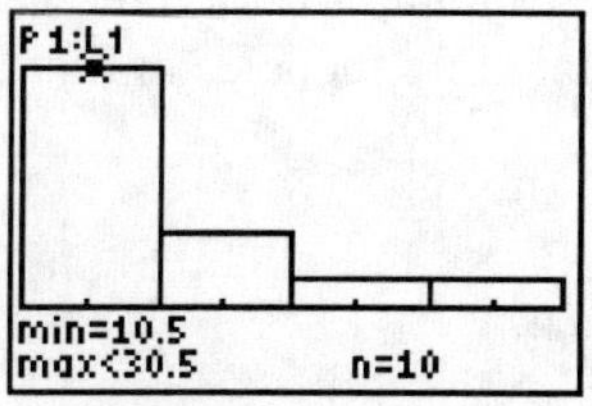

Graphing the histogram and then tracing should yield this screen.

3.

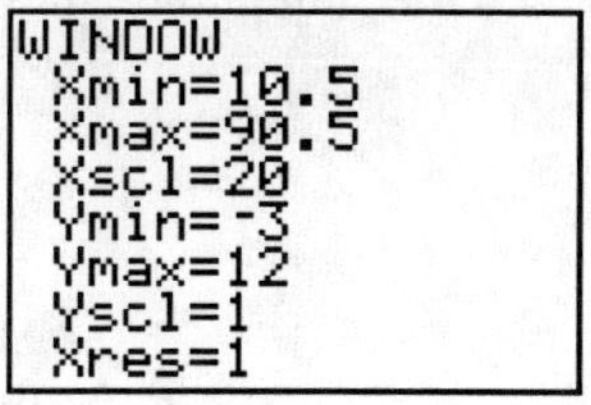

These are good window settings.

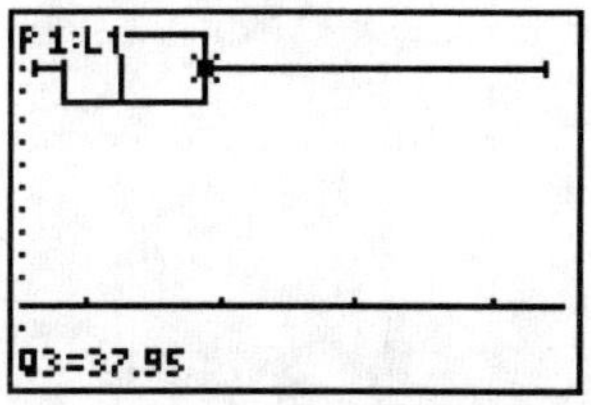

Graphing the boxplot and then tracing should yield this screen.

4.

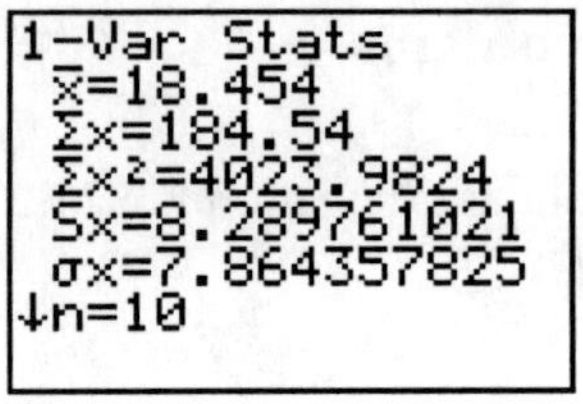

The mean is 18.454 inches, the standard deviation is 8.290 inches and the median is 17.08 inches. (Scroll down to see median.)

5.

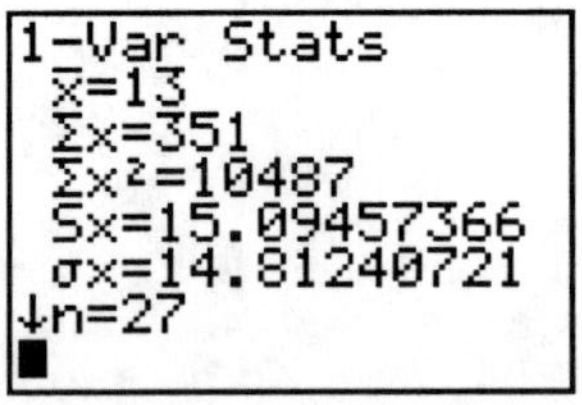

Remember to **Sort** the list in which you have entered the data. The mean is 13 movies per year, the median is 6 movies per year, and the mode is 5 movies per year. (Scroll down to see median.)

6.

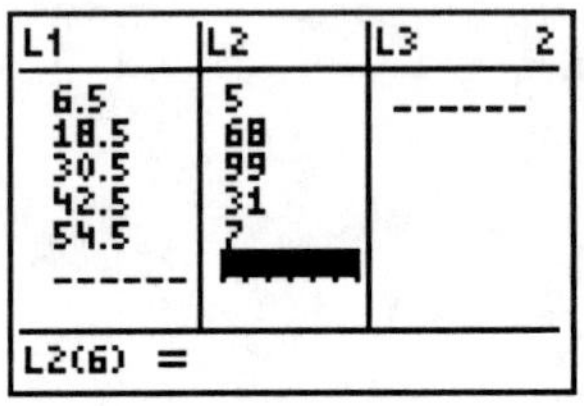

Enter class midpoints into List 1, and frequencies into List 2.

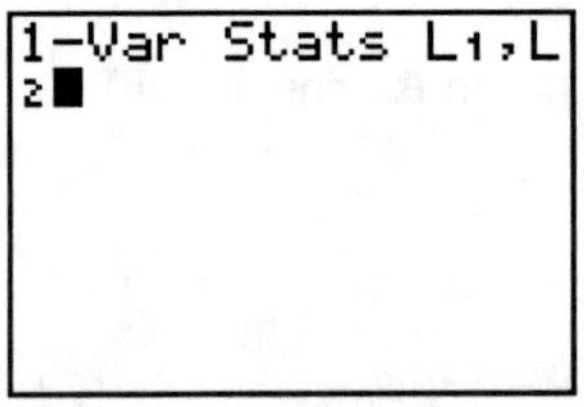

This command instructs the calculator to find the values of **x** in List 1 and the associated frequencies in List 2.

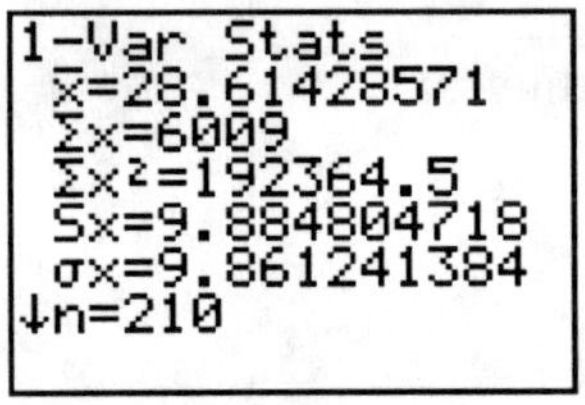

The mean is 28.61 minutes, and the standard deviation is 9.86 minutes.

Expected Value and Standard Deviation of a Discrete Probability Distribution

Example 1: Each day on a college campus parking tickets are issued. Let **x** = the number of parking tickets issued in a day. The following table shows the probability distribution of **x**.

No. of tickets issued in a day	5	6	7	8	9	10	11	12
Probability	.12	.14	.10	.20	.15	.16	.08	.05

Find the ***expected number*** of parking tickets for a day. (The expected number, or expected value, is the same as the <u>mean</u>.) Also, find the standard deviation.

[For **TI-82**, use program **PROBDIST**, as the **TI-82** will not accept decimals as frequencies. See page 85.]

FIRST: Enter **x** values in List 1, and P(**x**) values in List 2. Remember to clear old entries first and to check your data entry.

SECOND: Calculate the desired statistics.

Steps

STAT	
▶	Select **CALC**.
1	Select **1-Var Stats**.
2nd	Specify that the values
L1	of **x** are in **L1**,
,	and the associated probabilities
2nd	(relative frequencies)
L2	are in **L2**.
ENTER	

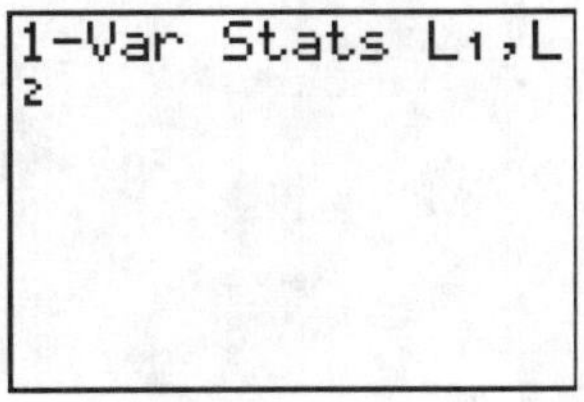

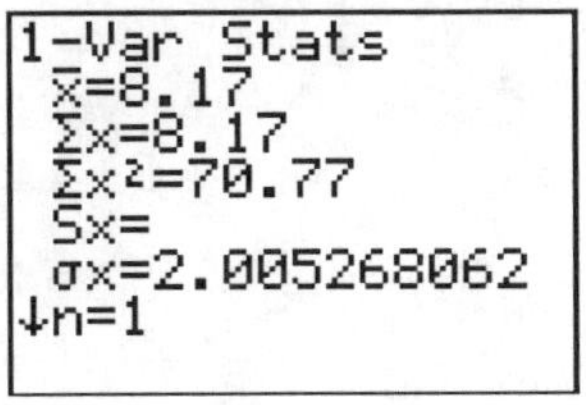

Solution: The expected number of tickets (mean) is 8.17. The standard deviation is 2.01.

Example 2: Suppose that you have purchased one $2 ticket for a lottery in which 10,000 tickets were sold. There is one first prize worth $5000, two second prizes worth $500 each, five third prizes worth $50 each and 100 fourth prizes worth $10 each. Find the mean of this distribution, that is, what is the "*expected winnings*?"

NOTE: First set up a table letting **x** = net winnings. Find **x** and P(**x**) for each outcome.

Outcomes	x	P(x)
first prize	4998	1/10,000
second prize	498	2/10,000
third prize	48	5/10,000
fourth prize	8	100/10,000
no prize	-2	9892/10,000

[For **TI-82**, use program **PROBDIST**, as the **TI-82** will not accept decimals as frequencies. See page 85.]

FIRST: Enter **x** values in List 1 and P(**x**) values in List 2. Remember to clear old entries first, and to check your data entry. Notice that when you type **1/10,000**, the calculator displays the scientific notation **1E-4**, which is $1 * 10^{-4}$.

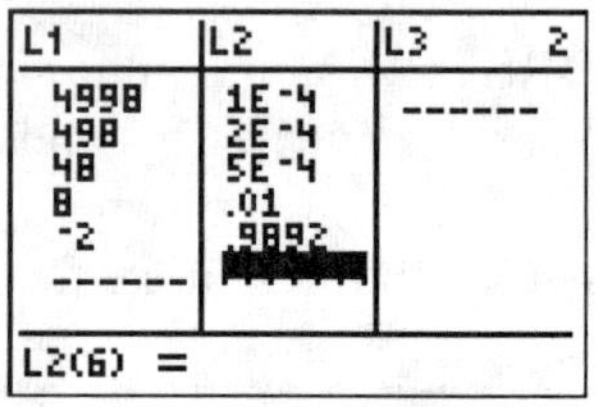

SECOND: Calculate the **1-Var Stats**.

Steps

STAT

▶ Select **CALC**.

1 Select **1-Var Stats**.

2nd **L1** Specify that the values of **x** are in **L1**,

, **2nd** **L2** and the associated probabilities are in **L2**.

ENTER

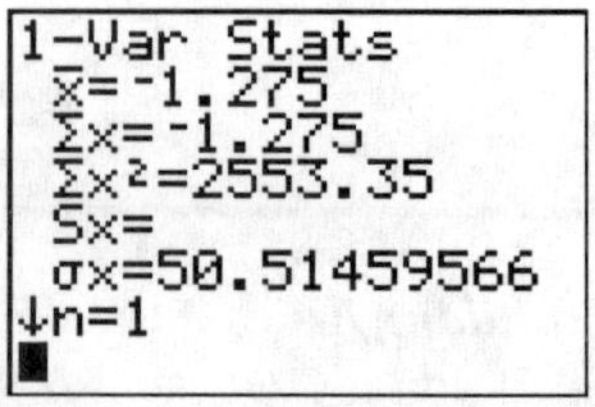

Solution: The "expected winnings" is really a loss of $1.275.

Mann: 5.3, 5.4

Binomial Probabilities

Example 1: Suppose a coin is tossed 10 times. Find the probability that "heads" come up exactly 6 times.

NOTE: For this problem the probability of success "**p**" is .5, the number of trials "**n**" is 10, and the number of successes "**x**" is 6.

[For **TI-82**, use the program **BINOML82**, as the **TI-82** does not have a Distribution key. See page 86.]

Steps	
2ND DISTR	Select the distribution menu.
0	Select **binompdf(**. This function computes the binomial probability of a single, specific number of successes.
10	Enter the number of trials.
,	(**n** = 10)
.5	Enter the probability of success on a single trial.
,	(**p** = .5)
6	Enter the number of successes.
)	(**x** = 6)
ENTER	

```
binompdf(10,.5,6
)
           .205078125
```

Solution: The probability of getting <u>exactly</u> 6 heads on 10 throws is .2051.

$P(\mathbf{x} = 6) = .2051$

(The program **BINOML83** for the **TI-83** makes this problem easier. See page 86.)

Example 2: Suppose that a coin is tossed 10 times. Find the probability that "heads" comes up at most 6 times.

NOTE: For this problem the probability of success "**p**" is .5, the number of trials "**n**" is 10, and the number of successes "**x**" is 0, 1, 2, 3, 4, 5, or 6.

[For **TI-82**, use the program **BINOML82**, as the **TI-82** does not have a Distribution key. See page 87.]

Steps

2nd **DISTR**	Select the distribution menu.
Scroll Down **ALPHA** **A**	Select **binomcdf(**. This function computes the cumulative probability starting with 0 successes.
10	Enter the number of trials. (**n** = 10)
, **.5**	Enter the probability of success on a single trial. (**p** = .5)
, **6** **)**	Enter the maximum number of successes. ($0 \le \mathbf{x} \le 6$)
ENTER	

```
binomcdf(10,.5,6
)
              .828125
```

Solution: The probability of getting at most 6 heads on 10 throws is .8281.

$P(0 \le \mathbf{x} \le 6) = .8281$

(The program **BINOML83** for the **TI-83** makes this problem easier. See page 87.)

Mann: 5.6

Example 3: The probability that a given basketball player makes any single free throw is .78. Find the probability that if this player shoots 200 free throws, he will make at least 160, but less than 180 free throws.

NOTE: For this problem the probability of success "**p**" is .78, the number of trials "**n**" is 200, and the number of successes "**x**" is 160, 161, 162, … 179.

[For **TI-82**, use the program **BINOML82**, as the **TI-82** does not have a Distribution key. See page 88.]

Steps

2nd **DISTR** **Scroll Down** **ALPHA**	Select the distribution menu.
A	Choose **binomcdf(**. This function computes the cumulative probability starting with 0 successes.
Enter the values	Enter the values shown to find $P(0 \le \mathbf{x} \le 179)$.

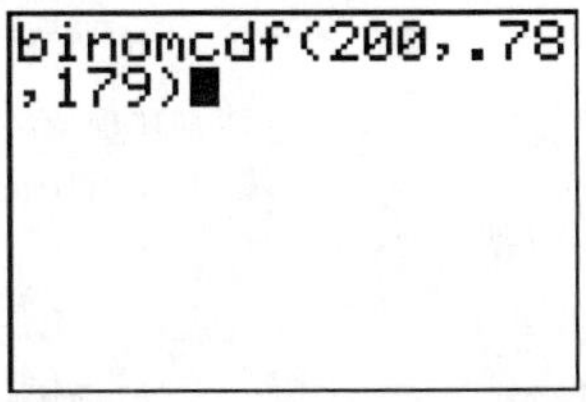

–	Subtract.
2nd **DISTR** **Scroll Down** **ALPHA**	Choose **binomcdf(**.
A	
Enter the values	Enter the values shown to find $P(0 \le \mathbf{x} \le 159)$.

ENTER	Compute $P(0 \le \mathbf{x} \le 179) - P(0 \le \mathbf{x} \le 159)$ which equals $P(160 \le \mathbf{x} \le 179)$, which equals $P(160 \le \mathbf{x} < 180)$.

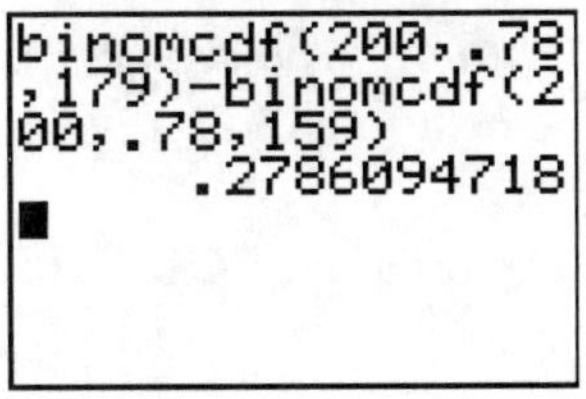

Solution: The probability that if this player shoots 200 free throws, he will make at least 160, but less that 180 free throws is .2786. $P(160 \le \mathbf{x} < 180) = .2786$

(The program **BINOML83** for the **TI-83** makes this problem easier. See page 88.)

Mann: 5.6

Poisson Probabilities

Example 1: A photocopy machine at a college breaks down an average of 8 times a month. Using the Poisson probability distribution, find the probability that during the next month the photocopy machine will break down exactly 9 times.

NOTE: In this problem the mean number of breakdowns per month "λ" is 8.

[For **TI-82**, use the program **POISSN82**, as the **TI-82** does not have a Distribution key. See page 90.]

Steps	
2nd DISTR	Select the distribution menu.
Scroll Down	
ALPHA	
B	Select **poissonpdf(**. This function calculates the Poisson probability for a single, specific value of **x**.
8	Enter the mean number of breakdowns per month. (λ = 8)
,	
9	Enter the specific number of breakdowns. (**x** = 9)
)	
ENTER	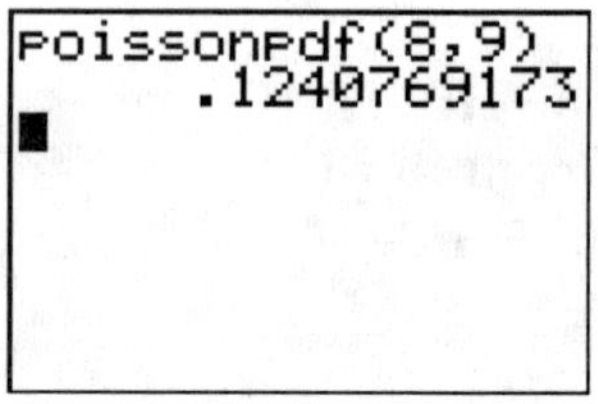

Solution: The probability that the photocopy machine will break down exactly 9 times in the next month is .1241. P(**x** = 9) = .1241

(The program **POISSN83** for the **TI-83** makes this problem easier. See page 90.)

Example 2: A photocopy machine at a college breaks down an average of 8 times a month. Using the Poisson probability distribution, find the probability that during the next month the photocopy machine will break down <u>less than</u> 8 times.

NOTE: In this problem the mean number of breakdowns per month "λ" is 8.

[For **TI-82**, use the program **POISSN82**, as the **TI-82** does not have a Distribution key. See page 91.]

Steps

2nd DISTR	Select the distribution menu.
Scroll Down **ALPHA** **C**	Select **poissoncdf(**. This function calculates the cumulative probability starting with 0 breakdowns.
8	Enter the mean number of breakdowns per month. (λ = 8)
,	
7	Enter the maximum number of breakdowns per month.
)	(0 ≤ **x** ≤ 7)
ENTER	

Solution: The probability that the photocopy machine will break down less than 8 times in the next month is .4530. P(**x** < 8) = P(0 ≤ **x** ≤ 7) = .4530

(The program **POISSN83** for the **TI-83** makes this problem easier. See page 91.)

Probabilities in a Normal Distribution

Example 1: If **z** has a standard normal distribution, find the probability that a value of **z** falls more than ½ of a standard deviation to the right of the mean.

[For **TI-82**, use the program **NORMAL82**, as the **TI-82** does not have a Distribution key. See page 92.]

NOTE: In the standard normal distribution, $\mu = 0$ and $\sigma = 1$.

Steps

Key	Action	Screen
2ND **DISTR**	Select the distribution menu.	
2	Select **normalcdf(**. This function computes the normal distribution probability between two bounds for a given mean and standard deviation.	
.5	Enter lower bound.	
,		
10^99	Enter upper bound. (**10^99** approximates ∞.)	
,		
0	Enter the mean.	
,		
1	Enter standard deviation.	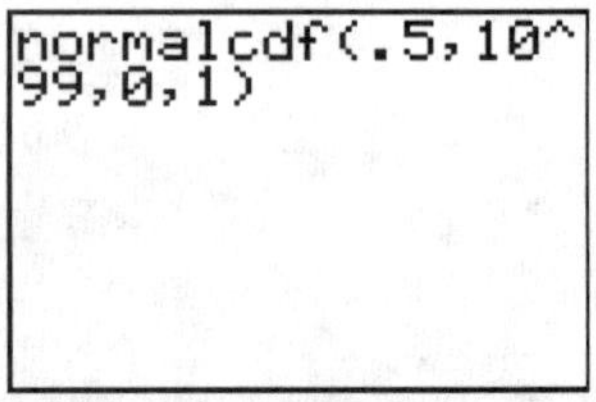
)		
ENTER		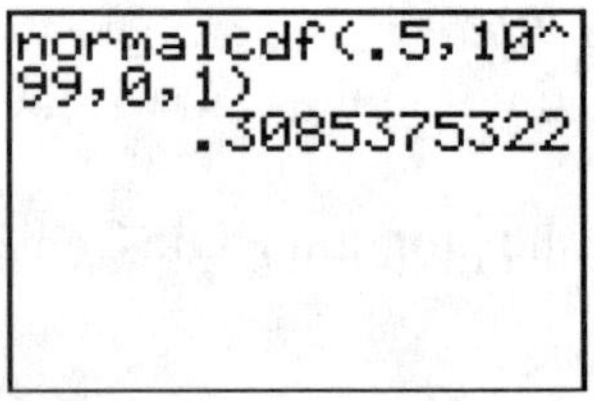

Solution: If **z** has the standard normal distribution, the probability that **z** is more than .5 standard deviation to the right of the mean is .3085. P(**z** >.5) = .3085

(The program **NORMAL83** for the **TI-83** makes this problem easier. See page 92.)

Example 2: A normal distribution has a mean of 10 and a standard deviation of 2. Find the probability that a randomly chosen value is less than 13.

[For **TI-82**, use the program **NORMAL82**, as the **TI-82** does not have a Distribution key. See page 93.]

Steps		
2ND DISTR	Select the distribution menu.	
2	Choose **normalcdf(**. This function computes the normal distribution probability between two bounds for a given mean and standard deviation.	
-10^99	Enter lower bound.	
,	(**-10^99** approximates -∞.)	
13	Enter the upper bound.	
,		
10	Enter the mean.	
,		
2	Enter the standard deviation.	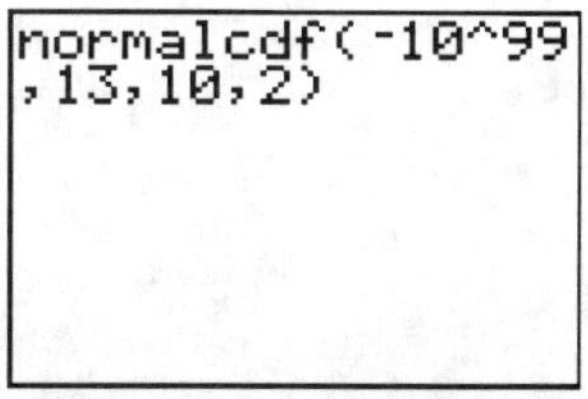
)		
ENTER		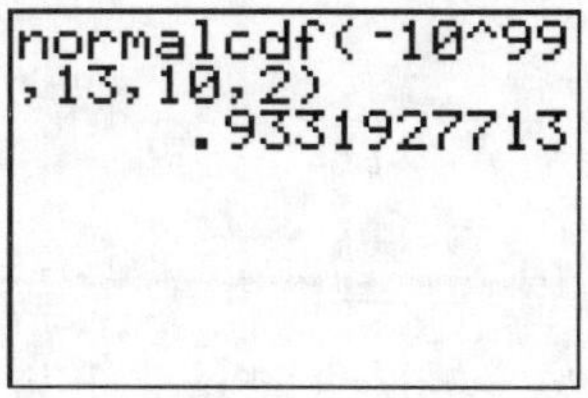

Solution: The probability that **x** is less than 13 is .9332. P(**x** < 13) = .9332

(The program **NORMAL83** for the **TI-83** makes this problem easier. See page 93.)

Example 3: The normal distribution of "IQ" in the population has a mean of 100, and a standard deviation of 15. What is the probability that a randomly selected individual will have an IQ between 110 and 120?

[For **TI-82**, use the program **NORMAL82**, as the **TI-82** does not have a Distribution key. See page 94.]

Steps

2ND **DISTR**	Select the distribution menu.
2	Choose **normalcdf(**. This function computes the normal distribution probability between two bounds for a given mean and standard deviation.
110	Enter the lower bound.
, **120**	Enter the upper bound.
, **100**	Enter the mean.
, **15** **)**	Enter the standard deviation.
ENTER	

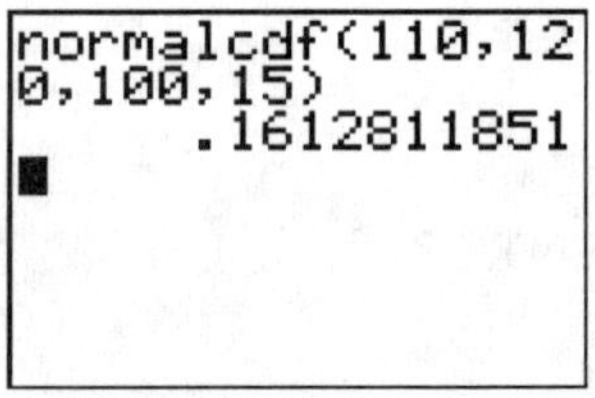

Solution: The probability is .1613 that the IQ score of a randomly selected individual is between 110 and 120. $P(110 < \mathbf{x} < 120) = .1613$

(The program **NORMAL83** for the **TI-83** makes this problem easier. See page 94.)

Example 4: Draw a graph that illustrates P(110 < **x** < 120) in a normal distribution with a mean of 100 and a standard deviation of 15.

[For **TI-82**, use the program **NORMAL82**, as the **TI-82** does not have a Distribution key. See page 94.]

FIRST: <u>Turn</u> <u>off</u> the graphs of any functions or **STAT PLOT**s currently turned on.

Steps

VARS	[On **TI-82**, use **2nd** **Y-VARS**.]
▶	Select **Y-VARS**.
4 [**5** on **TI-82**]	Select **On/Off....**
2	Select **FnOff**.
ENTER	
2ND	
STAT PLOT	
4	Select **PlotsOff**.
ENTER	**Note:** Any old functions will be retained in your calculator, but they <u>will</u> <u>not</u> <u>be</u> <u>graphed</u> until you issue the command **FnOn**. Use the **PlotsOn** command to revive plots.

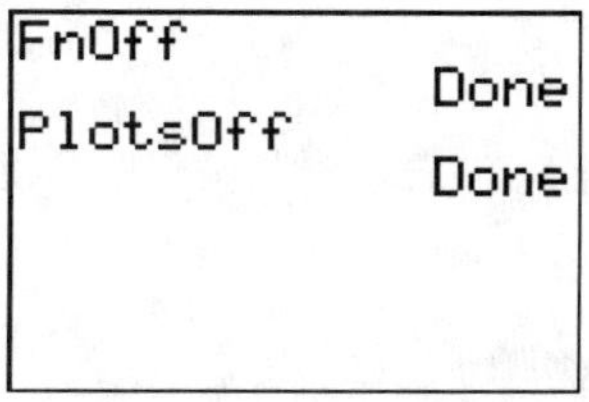

SECOND: Set the window so that the graph will appear on the screen.

Steps

WINDOW	The **WINDOW** key is located just beneath the screen. This key is used to set up the domain, range and scale of a graph.
Enter settings as shown	Set **Xmin** at 3 standard deviations to the left of the mean (55 = 100-3∗15). Set **Xmax** at 3 standard deviations to the right of the mean (145 = 100+3∗15). Set the scale, **Xscl**, at one standard deviation, 15. Set **Ymin** = -.2/σ. (-.2/15) Set **Ymax** = .5/σ. (.5/15)

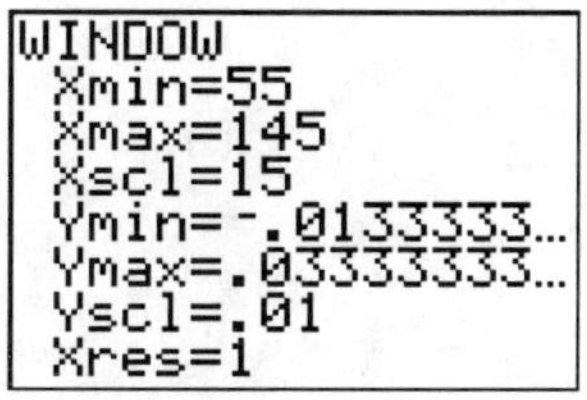

THIRD: Draw the graph.

Steps	
2nd DISTR	Select the distribution menu.
▶	Select **DRAW**.
1	Select **ShadeNorm(**. This function draws a normal distribution, shaded from the lower bound to the upper bound.
110	Enter the lower bound.
,	
120	Enter the upper bound.
,	
100	Enter the mean.
,	
15	Enter standard deviation.
)	
ENTER	

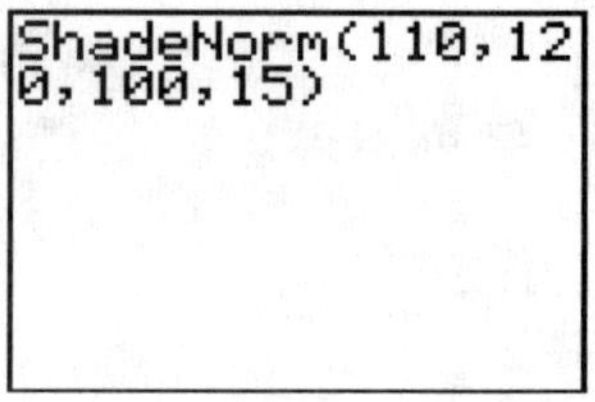

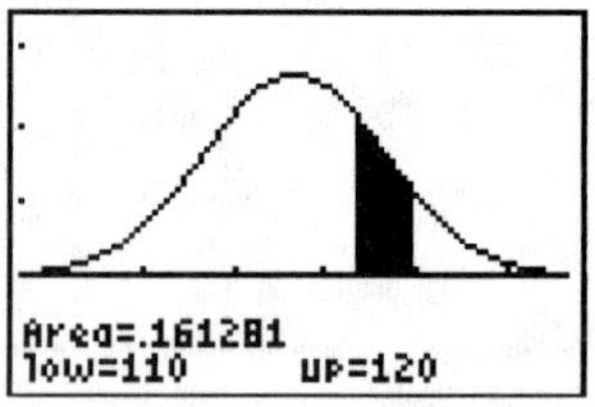

(The program **NORMAL83** for the **TI-83** makes this problem easier. See page 94.)

Finding z in a Normal Distribution

Example 1: Find the **z** value for a standard normal distribution if the area in the left tail is .05.

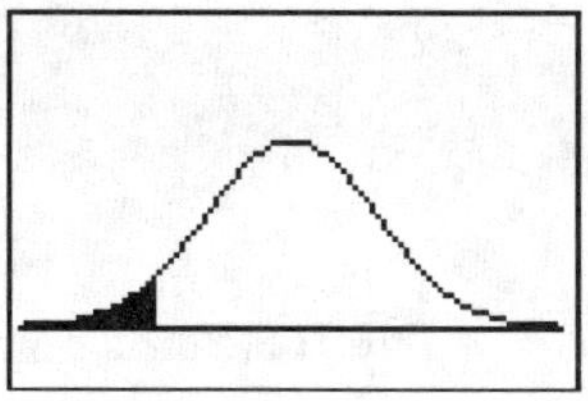

[For **TI-82**, use the program **INVNOR82**, as the **TI-82** does not have a Distribution key. See page 95.]

Steps	
2nd DISTR	Select the distribution menu.
3	Select **invNorm(**. This inverse function pairs the area under the curve, in the left tail, with the corresponding **z** value.
.05	Enter the area in the left tail.
,	
0	Enter the mean.
,	
1	Enter the standard deviation.
)	
ENTER	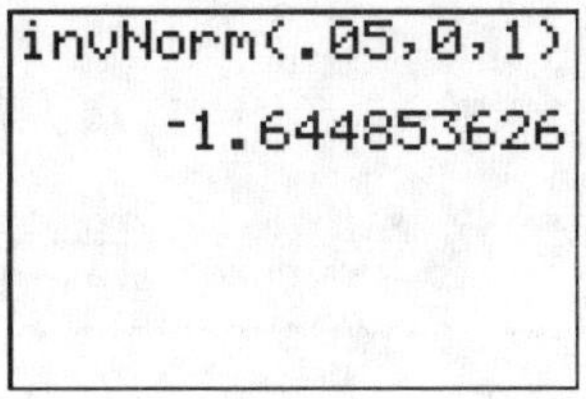

Solution: The **z** value = -1.645, when the area in the left tail of the standard normal distribution is .05. P(**z** < -1.645) = .05

(The program **INVNOR83** for the **TI-83** makes this problem easier. See page 95.)

Example2: Find the **z** value for a standard normal distribution if the area in the right tail is 0.125.

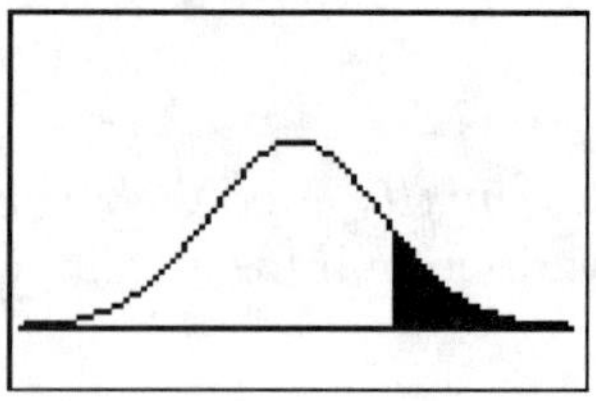

[For **TI-82**, use the program **INVNOR82**, as the **TI-82** does not have a Distribution key. See page 96.]

Steps	
2nd DISTR	Select the distribution menu.
3	Select **invNorm(**. This inverse function pairs the area under the curve, in the left tail, with the corresponding **z** value.
1−.125	Enter the area in the left tail.
,	
0	Enter the mean.
,	
1	Enter the standard deviation.
)	
ENTER	

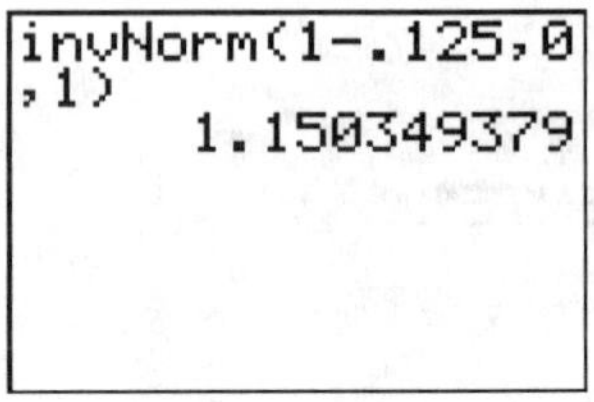

Solution: The **z** value = 1.15, when the area in the right tail of the standard normal distribution is .125. P(**z** >1.15) = .125

(The program **INVNOR83** for the **TI-83** makes this problem easier. See page 96.)

Example 3: Suppose that the height of adult males has a normal distribution with a mean of 70 inches and a standard deviation of 3 inches. How high should a door be so that 98% of adult males can pass through without ducking?

[For **TI-82**, use the program **INVNOR82**, as the **TI-82** does not have a Distribution key. See page 97.]

Steps	
2nd DISTR	Select the distribution menu.
3	Select **invNorm(**. This inverse function pairs the area under the curve, in the <u>left</u> tail, with the corresponding **z** value.
.98	Enter the area in the <u>left</u> tail.
,	
70	Enter the mean.
,	
3	Enter the standard deviation.
)	
ENTER	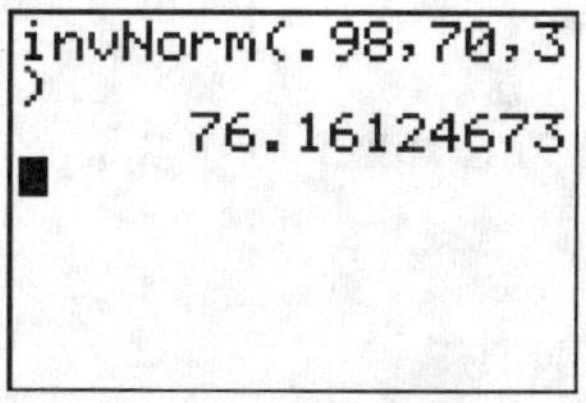

Solution: The height of a door should be 76.16 inches to permit 98% of adult men to pass through without ducking. $P(\mathbf{x} < 76.16) = .98$

(The program **INVNOR83** for the **TI-83** makes this problem easier. See page 97.)

Probabilities in the Normal Distribution of $\bar{x}$

Example: If the ages of all statistics students are distributed normally with a mean of 22.7 years and a standard deviation of 4.2 years, what is the probably that, for a sample of 35 statistics students, the mean age would be between 20 and 22 years?

[For **TI-82**, use the program **NORMAL82**, as the **TI-82** does not have a Distribution key. See page 93.]

Steps	
2ND DISTR	Select the distribution menu.
2	Select **normalcdf(**. This function computes the normal distribution probability between two bounds for a given mean and standard deviation.
20	Enter the lower bound.
,	
22	Enter the upper bound.
,	
22.7	Enter the mean.
,	
4.2/√(35)	Enter the standard deviation of $\bar{x}$.
)	
ENTER	

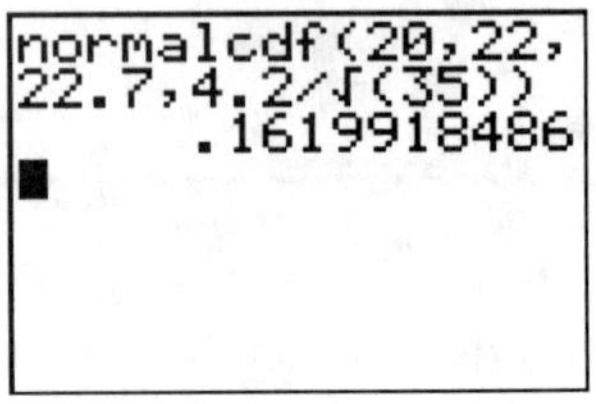

Solution: The probability that the mean age of the sample of 35 statistics students is between 20 and 22 years is .1620. $P(20 < \bar{x} < 22) = .1620$

(The program **NORMAL83** for the **TI-83** makes this problem easier. See page 93.)

Supplementary Problems

1. The following table gives the probability distribution for the number of Super Giant Sundaes sold daily at an ice cream store.

x = # of Sundaes	0	1	2	3	4	5
P(**x**)	.1	.2	.4	.15	.1	.05

Find the expected number of Super Giant Sundaes sold on a given day.

2. 63% of high school juniors take the SAT in preparation to apply to colleges. What is the probability that in a sample of 50 high school juniors, more than 40 of them take the SAT?

3. A police officer writes an average of 6.5 citations a day. Find the Poisson probability that this officer writes between 30 and 35 (inclusive) citations in a five day work week.

4. For the standard normal distribution, find P(-1< **z** <1).

5. In a company, the commute time for employees is normally distributed with a mean of 56.7 minutes and a standard deviation of 18.9 minutes. What percent of the employees commute for more than an hour?

6. Find the value of **z** so that the area under the standard normal curve to the left of **z** is .1765.

7. An author has been told that his textbook is longer than 95% of the textbooks of its type. If these textbook lengths are normally distributed with a mean of 753.2 pages and a standard deviation of 30.6 pages, what is the number of pages in this author's text?

8. Suppose that the cost of tennis shoes is normally distributed with a mean of $48.35 and a standard deviation of $13.52. What is the probability that for a random sample of 30 tennis shoes, the mean price is less than $50?

Solutions for Supplementary Problems

1.

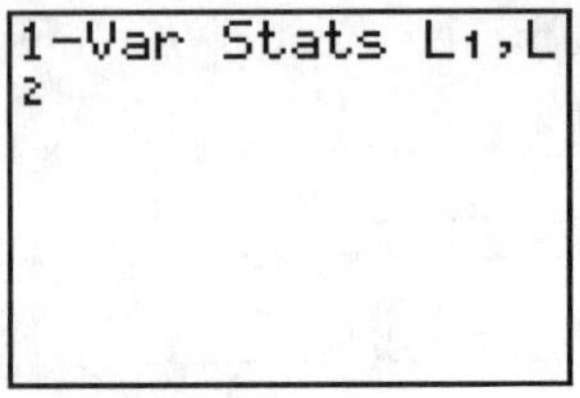

Enter the values of **x** into **L1** and P(**x**) into **L2**, then use the command **STATS>CALC>1-Var Stats** with argument **L1,L2**.

[For **TI-82**, use program **PROBDIST**. See page 85.]

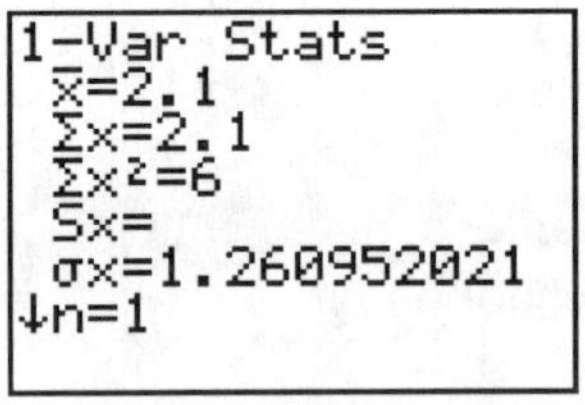

The expected number of sundaes sold per day is 2.1.

2.

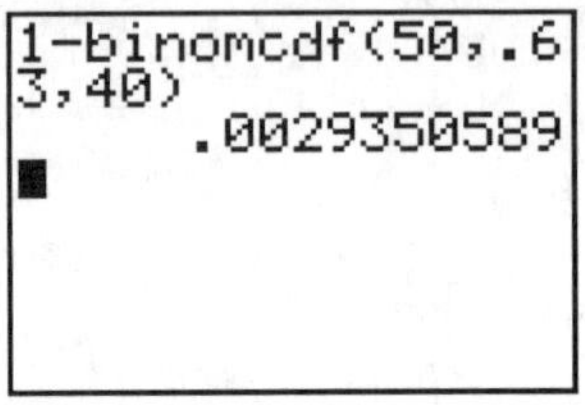

The probability is .0029 that in a sample of 50 students <u>more</u> than 40 of them will take the SAT.
$P(\mathbf{x} > 40) = P(\mathbf{x} \geq 41) = .0029$

[For **TI-82**, use program **BINOML82**. See page 86.]

(For **TI-83**, try program **BINOML83**. See page 86.)

3.

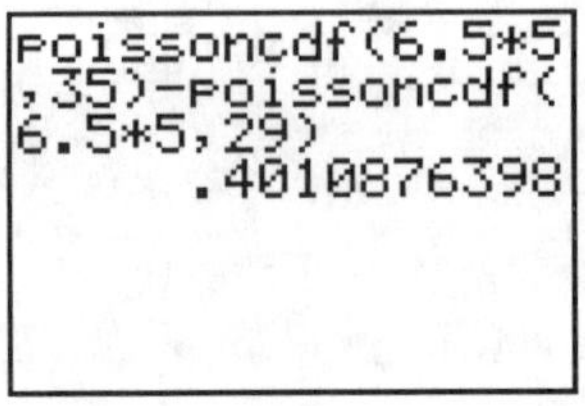

λ is 6.5*5. The Poisson probability that this officer writes between 30 and 35 (inclusive) citations in a <u>five day</u> work <u>week</u> is .4011.

[For **TI-82**, use program **POISSN82**. See page 90.]

(For **TI-83**, try program **POISSN83**. See page 90.)

4.

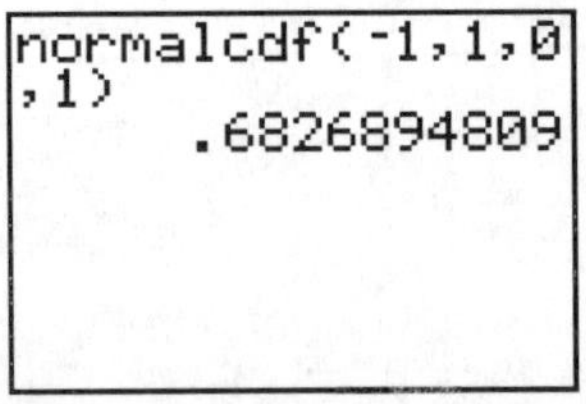

$P(-1 < \mathbf{z} < 1) = .6827$

[For **TI-82**, use program **NORMAL82**. See page 92.]

(For **TI-83**, try program **NORMAL83**. See page 92.)

5.

```
normalcdf(60,10^
99,56.7,18.9)
     .4306957284
```

The percent of employees that commute for more than an hour is 43.07%.

[For **TI-82**, use program **NORMAL82**. See page 92.]

(For **TI-83**, try program **NORMAL83**. See page 92.)

6.

```
invNorm(.1765,0,
1)
    -.9287859976
■
```

If the area under the normal curve to the left of **z** is .1765, then **z** = -.929.
P(**z** < -.929) = .1765

[For **TI-82**, use program **INVNOR82**. See page 95.]

(For **TI-83**, try program **INVNOR83**. See page 95.)

7.

```
invNorm(.95,753.
2,30.6)
      803.532521
```

The author's text is at least 804 pages.

[For **TI-82**, use program **INVNOR82**. See page 95.]

(For **TI-83**, try program **INVNOR83**. See page 95.)

8.

```
normalcdf(-10^99
,50,48.35,13.52/
√(30))
     .7480763622
■
```

The probability that for a sample of 30 pairs of tennis shoes the sample mean is than $50 is .7481.
P($\bar{x}$ < 50) = .7481

[For **TI-82**, use program **NORMAL82**. See page 92.]

```
normalcdf(0,50,4
8.35,13.52/√(30)
)
     .7480763622
■
```

Note: In this particular problem, 0 can be used for the lower bound, instead of -10^99, because 0 is many standard deviations to the left of the mean (and coincidentally the lowest possible price).

(For **TI-83**, try program **NORMAL83**. See page 92.)

Confidence Interval for the Mean
Using Summary Statistics and Normal Distribution

Example: A resident is trying to predict the mean speed at which cars are driven on her street. She borrows a radar gun and after gathering data on 250 cars, she finds a mean speed of 39.6 mph with a standard deviation of 7.4 mph. Find the 98% confidence interval for the mean speed of all cars driven on her street.

NOTE: With this large sample, our confidence interval will be constructed using a normal distribution where we approximate σ by **s**. Compare with the example on the next page.

FIRST: Organize your data, noting the summary statistics:

$\sigma \approx \mathbf{s} = 7.4$ $\quad \bar{\mathbf{x}} = 39.6$ $\quad \mathbf{n} = 250$ $\quad$ Confidence Level = .98

SECOND: Calculate the confidence interval from these summary statistics.

[For **TI-82**, use program **ONEPOPCI**. See page 100.]

Steps

STAT	
▸▸	Select **TESTS**.
7	Select **Z Interval**.
Stats	Move cursor to highlight **Stats**.
ENTER	
Enter the values	Move the cursor and enter the values shown, replacing old entries. Press **ENTER** after each entry. Approximate σ with **s**.

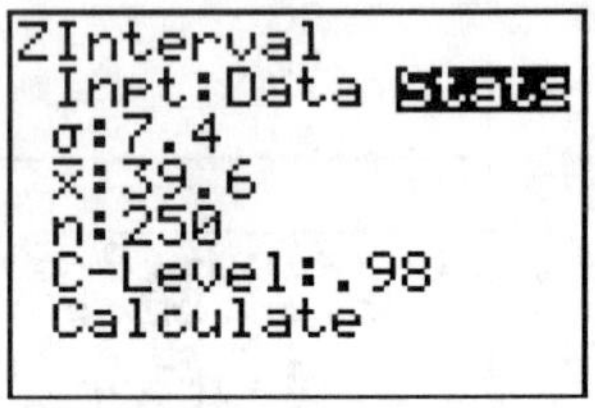

Calculate	Highlight **Calculate**.
ENTER	

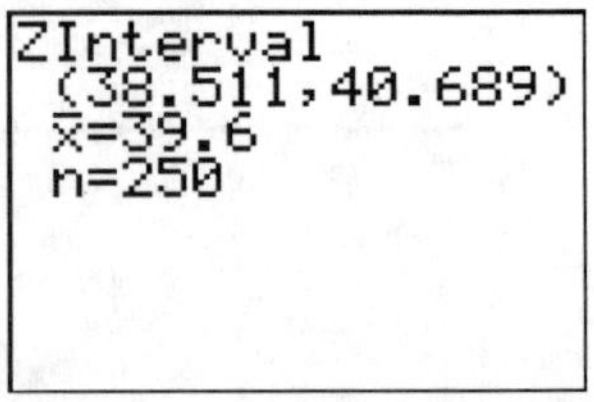

Solution: The woman is 98% confident that the mean speed of all cars driven on her street is between 38.511 mph and 40.689 mph.

Confidence Interval for the Mean
Using Summary Statistics and *t* Distribution

Example: A resident is trying to predict the mean speed at which cars are driven on her street. She borrows a radar gun and after gathering data on 250 cars, she finds a mean speed of 39.6 mph with a standard deviation of 7.4 mph. Find the 98% confidence interval for the mean speed of all cars driven on her street.

NOTE: Even though our sample is large, as we only have **s**, and not σ, the confidence interval will be constructed using the ***t*** distribution. Compare with the example on the previous page.

[The **TI-82** can not use a ***t*** distribution for large samples.]

FIRST: Organize your data, noting the summary statistics:

$\bar{\mathbf{x}} = 39.6$ **s** = 7.4   **n** = 250   Confidence Level = .98

SECOND: Calculate the confidence interval from these summary statistics.

Steps

STAT	
▶ ▶	Select **TESTS**.
8	Select **T Interval**.
Stats	Move cursor to highlight **Stats**.
ENTER	
Enter the values	Enter the values shown, writing over old entries. Press **ENTER** after each entry.

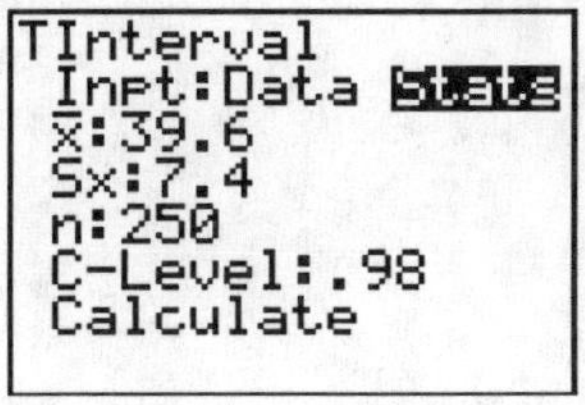

Calculate	Highlight **Calculate**.
ENTER	

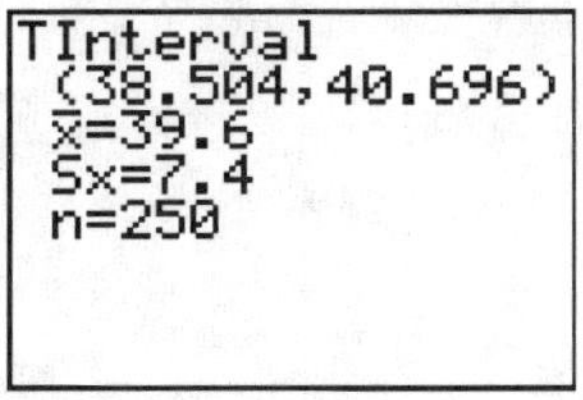

Solution: The woman is 98% confident that the mean speed of all cars driven on her street is between 38.504 mph and 40.696 mph.

Confidence Interval for the Mean Using Raw Data and *t* Distribution

Example: A teacher would like to predict how much time his students spend on homework each evening which he assumes to be normally distributed. He samples 12 random students and gets the following responses (in hours).

1 2.5 3 4.5 6 1.5 .75 3.2 6.1 5 3 2.5

Find the 90% confidence interval to predict the mean amount of time students in general spend on homework each evening.

FIRST: Clear any old data from List 1. Enter the data above into List 1. Double check your data entry.

SECOND: Calculate your confidence interval from these "raw data." You do not need to first find the summary statistics from this sample before proceeding.

[For **TI-82**, use program **ONEPOPCI**. See page 100.]

Steps

STAT		
▸▸	Select **TESTS**.	
8	Select **T Interval**.	
Data	Move cursor to highlight **Data**.	
ENTER		
Enter the values	Enter the values shown, writing over old entries. Press **ENTER** after each entry.	
Calculate **ENTER**	Highlight **Calculate**.	

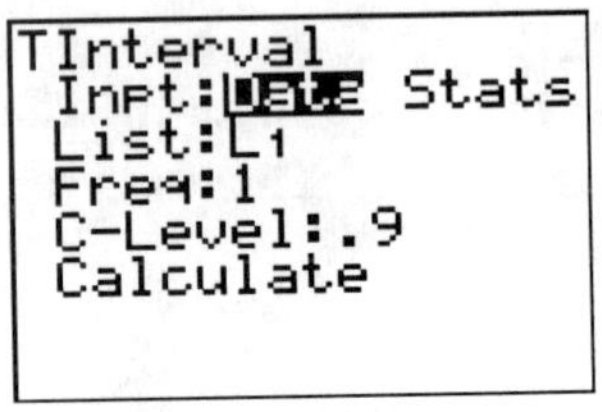

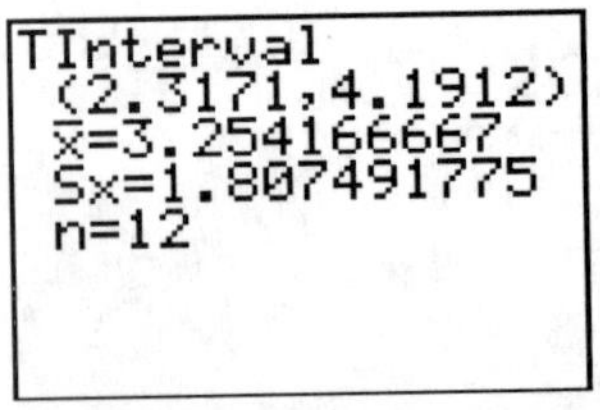

Solution: The teacher is 90% confident that the mean study time for all students is between 2.317 and 4.191 hours per night.

Confidence Interval for the Population Proportion Using Raw Data and Normal Distribution

Example: 600 persons were asked if they enjoyed the TV coverage of the 1996 Olympics. 438 respondents said that they did enjoy it. Find the 95% confidence interval for the proportion of the entire population, **p**, which enjoyed the games.

NOTE: As $n\hat{p} > 5$, and $n\hat{q} > 5$, the normal distribution is used to construct the confidence interval.

FIRST: Organize your data, noting the summary statistics:

x = 438 **n** = 600 ($\hat{p}$ = .73, and $\hat{q}$ = .27) Confidence Level = .95

SECOND: Calculate the confidence interval.

[For **TI-82**, use program **ONEPOPCI**. See page 100.]

Steps

STAT	
▶▶	Select **TESTS**.
Scroll Down	
Alpha	Select **1-PropZInt....**
A	
Enter the values	Enter the values shown, writing over old entries. Press **ENTER** after each entry.

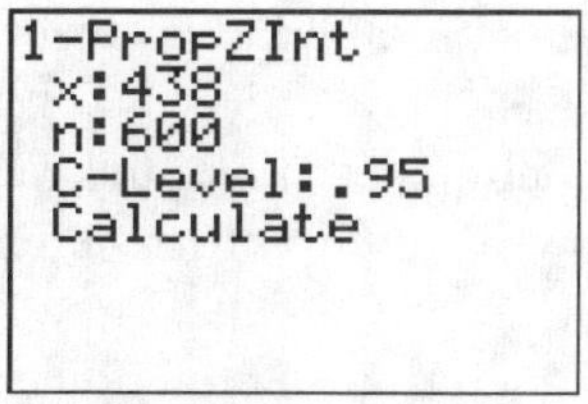

Calculate	Highlight **Calculate**.
ENTER	

Solution: We can be 95% confident that the proportion of the population which enjoyed the TV coverage of the Olympics is between .694 and .766.

NOTE: To calculate the confidence interval for p using $\hat{p}$, when x is not known: remember that x = $n\hat{p}$. Furthermore, on the calculator, x must be an integer. If necessary, round x to the nearest integer.

Hypothesis Test about the Mean
Using Summary Statistics and Normal Distribution

Example: Has there been grade inflation? In 1980, the mean grade point average of all college students at the end of their freshman year was 2.97. In a recent study of 45 randomly selected college students, it was found that their mean GPA at the end of the freshman year was 3.16 with a standard deviation of .54. Test at the 2.5% significance level if the mean GPA today is higher than 2.97.

NOTE: This is a large sample, so use the normal distribution. The critical value is $\mathbf{z} = 1.960$. The rejection region is to the right of $\mathbf{z} = 1.960$.

FIRST: Set up the problem.

$H_0: \mu = 2.97 \qquad H_1: \mu > 2.97$

SECOND: Calculate the p-value and test statistic.

[For **TI-82**, use program **ONEPOPHT**. See page 100.]

Steps

STAT	
▶▶	Select **TESTS**.
1	Select **Z-TEST....**
Stats	Move cursor to highlight **Stats**.
ENTER	
Enter the values	Enter the values shown, writing over old entries. Press **ENTER** after each entry.
▼	Scroll down and highlight the type of alternative hypothesis: **>μ_0**.
>μ_0	
ENTER	
Calculate	Highlight **Calculate**.
ENTER	

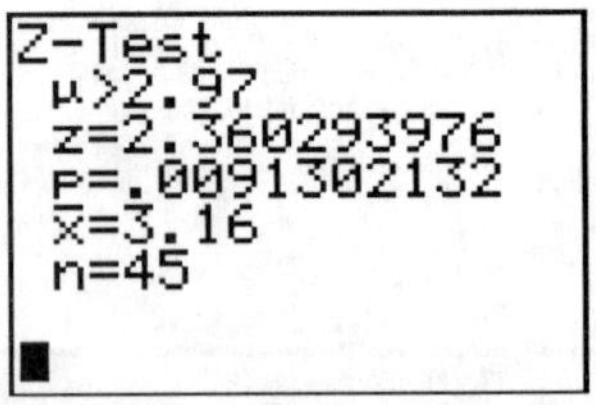

Decision: Reject H_0. The significance level = .025, and the p-value = .009. The critical value = 1.96, and the test statistic = 2.360. There is sufficient evidence to conclude that the mean GPA of current college students at the end of their freshman year is greater that 2.97.

Hypothesis Test about the Mean Using Raw Data and *t* Distribution

Example: Today, how do dogs rate as pets? For many decades Americans, when rating dogs as pets on a 10 point scale, have given them normally distributed scores with a mean score of 7.6. Has there been a change of opinion? In a recent study, a randomly selected sample of 17 Americans gave the following ratings of dogs as pets. Test at the 5% significance level if the mean rating today is lower than 7.6.

7 6 5 10 8 9 4 5 6 5 9 7 6 10 8 6 7

NOTE: This is a small sample, so use the t distribution. At 16 degrees of freedom, the critical value is $t = -1.746$. The rejection region is to the left of $t = -1.746$.

FIRST: Set up the problem. $H_0: \mu = 7.6$ $H_1: \mu < 7.6$

SECOND: Clear any old data from List 1. Enter the data above into List 1. Double check your data entry.

THIRD: Calculate the p-value and test statistic.

[For **TI-82**, use program **ONEPOPHT**. See page 100.]

Steps

STAT	
▶ ▶	Select **TESTS**.
2	Select **T-TEST....**
Data	Move cursor to highlight **Data**.
ENTER	
Enter the values	Enter the values shown, writing over old entries. Press **ENTER** after each entry.
▼	Scroll down and highlight the type of
$<\mu_0$	alternative hypothesis: $<\mu_0$.
ENTER	
Calculate	Highlight **Calculate**.
ENTER	

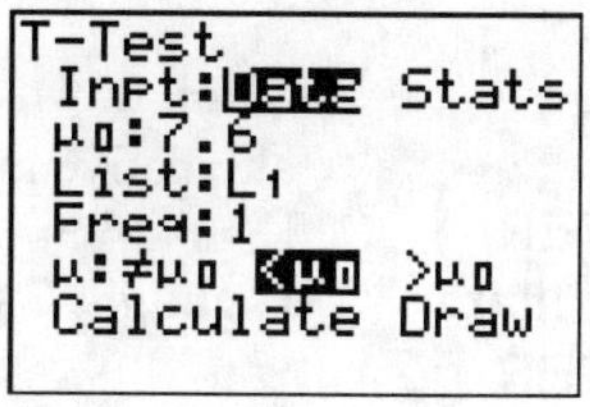

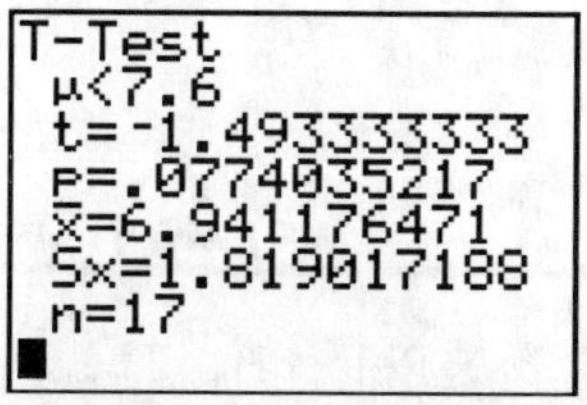

Decision: Do not reject H_0. The significance level = .05, and the p-value = .0774. The critical value = -1.746, and the test statistic = -1.493. There is not sufficient evidence to conclude that today dogs are rated lower as pets than they were in the past.

Mann: 9.4

Hypothesis Test about the Population Proportion Using Raw Data and Normal Distribution

Example: Suppose that 43% of adults in the country smoke tobacco. It is believed that smoking is less popular on the West Coast. To test this, a survey was taken of West Coast residents and 295 of 780 respondents reported that they used tobacco. Test at the 2% significance level if the proportion of adults on the West Coast who smoke is different from that of the country as a whole.

NOTE: This is a large sample as $\mathbf{n\hat{p}} > 5$, and $\mathbf{n\hat{q}} > 5$. The normal distribution will be used for this two-tail test. The critical values of **z** are 2.326 and -2.326. The rejection region is to the outside of these values.

FIRST: Set up the problem. $H_0: \mathbf{p} = .43$ $H_1: \mathbf{p} \neq .43$

SECOND: Organize your data.

$$\mathbf{x} = 295 \qquad \mathbf{n} = 780 \qquad \left(\hat{\mathbf{p}} = \frac{\mathbf{x}}{\mathbf{n}} = \frac{295}{780} = .3782\right)$$

THIRD: Calculate the p-value and test statistic.

[For **TI-82**, use program **ONEPOPHT**. See page 100.]

Steps

STAT	
▶ ▶	Select **TESTS**.
5	Select **T1-PropZTest...**.
Enter the values	Enter the values shown, writing over old entries. Press **ENTER** after each entry.
▼	Scroll down and highlight the type of alternative hypothesis: $\neq \mathbf{p_0}$.
$\neq \mathbf{p_0}$	
ENTER	
Calculate	Highlight **Calculate**.
ENTER	

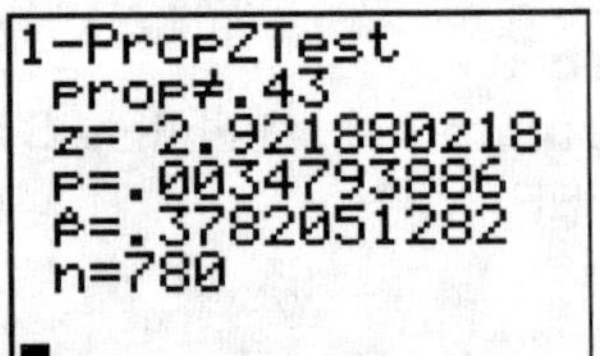

Decision: Reject H_0. The significance level = .02, and the p-value = .0035. The relevant critical value = -2.326, and the test statistic = -2.922. There is sufficient evidence to conclude that the proportion of smokers on the West Coast is different from that in the rest of the country.

Supplementary Problems

1. A random sample of 200 checking accounts at a bank showed a mean balance of $685 with a standard deviation of $85. Construct the 90% confidence interval for the mean balance of all checking accounts at this bank.

2. Fifteen college students in a random sample were asked how much they spent for lunch at the student cafeteria on a particular day. Their responses are shown below.

$4.12	5.35	1.26	4.38	5.01	3.98	3.74	4.25
5.69	3.78	2.56	4.98	6.89	5.33	3.22	

   Construct the 95% confidence interval for the mean amount spent on lunch by all students at this school.

3. Employees conducting a phone survey reported that of the 500 calls they made, 360 resulted in people being rude or hanging up on then. Construct a 98% confidence interval for the proportion of all phone survey respondents who are rude or hang up.

4. A college reports that 40% of its graduates get accepted at graduate school. In a sample of 680 graduates from this school, it was found that 250 had been accepted into graduate school. At the 5% significance level, can you conclude that the proportion of graduates from this school which get accepted at graduate school is less than 40% ?

5. A grocery store claims that the average time customers spend waiting in line at its store is 1.5 minutes. A random sample of 12 shoppers from this store showed the following waiting times (in minutes).

   1.25 2.5 5.75 3 4.5 3.25 5 .75 .25 .5 2.5 3.5

   At the 2.5% significance level, can you conclude that the waiting time is greater than 1.5 minutes?

6. An employer at a fast food establishment claims that the average length of employment for his employees is 6 months. In a random sample of 100 employees from this business, it was found that the mean length of employment was 5.6 months with a standard deviation of 3 months. At the 10% significance level, can you conclude that the mean time of employment at this establishment is different from 6 months?

Solutions for Supplementary Problems

1a.

```
ZInterval
 (675.11,694.89)
 x̄=685
 n=200
```

The 90% confidence interval for the mean balance of all checking accounts at this bank is \$675.11 to \$694.89. **Note:** This solution assumes you used a normal distribution because the sample was <u>large</u>.

[For **TI-82**, use program **ONEPOPCI**. See page 100.]

1b.

```
TInterval
 (675.07,694.93)
 x̄=685
 Sx=85
 n=200
```

The 90% confidence interval for the mean balance of all checking accounts at this bank is \$675.07 to \$694.93. **Note:** This solution assumes you used a t distribution because sigma was <u>unknown</u>.

[For **TI-82**, use program **ONEPOPCI**. See page 100.]

2.

```
TInterval
 (3.549,5.0564)
 x̄=4.302666667
 Sx=1.361020975
 n=15
```

The 95% confidence interval for the mean amount spent on lunch for all students at this school is \$3.55 to \$5.06.

[For **TI-82**, use program **ONEPOPCI**. See page 100.]

3.

```
1-PropZInt
 (.67329,.76671)
 p̂=.72
 n=500
```

The 98% confidence interval for the proportion of all telephone survey respondents who are rude or hang up is .673 to .767 or 67.3% to 76.7%.

[For **TI-82**, use program **ONEPOPCI**. See page 100.]

4.

```
1-PropZTest
 prop<.4
 z=-1.7221168
 p=.0425241445
 p̂=.3676470588
 n=680
```

H_0: $\mathbf{p} = .4$ $\quad$ H_1: $\mathbf{p} < .4$
The critical value is -1.645 and the test statistic is -1.722, and $\alpha = .05$, while the p-value = .0425. The decision is to reject H_0. At the 5% significance level, we can conclude that the proportion of graduates from this college which gets accepted at graduate school is less than 40%.

[For **TI-82**, use program **ONEPOPHT**. See page 100.]

5.

```
T-Test
 μ>1.5
 t=2.364130089
 p=.0187695431
 x̄=2.729166667
 Sx=1.801067655
 n=12
```

$H_0: \mu = 1.5$ $H_1: \mu > 1.5$

The critical value is 2.201 and the test statistic is 2.364. $\alpha = .025$, while the p-value = .0188. The decision is to reject H_0. At the 2.5% significance level, we can conclude that the mean waiting time at this store is greater than 1.5 minutes.

[For **TI-82**, use program **ONEPOPHT**. See page 100.]

6a.

```
Z-Test
 μ≠6
 z=-1.333333333
 p=.1824225637
 x̄=5.6
 n=100
```

$H_0: \mu = 6$ $H_1: \mu \neq 6$

The critical values of this two-tailed test are ± 1.645. The test statistic is -1.333. $\alpha = .10$, while the p-value = .1824. The decision is to not reject H_0. At the 10% significance level, we cannot conclude that the mean time of employment is different from 6 months.

Note: This solution assumes you used a normal distribution because the sample was large.

[For **TI-82**, use program **ONEPOPHT**. See page 100.]

6b.

```
T-Test
 μ≠6
 t=-1.333333333
 p=.1854823499
 x̄=5.6
 Sx=3
 n=100
```

$H_0: \mu = 6$ $H_1: \mu \neq 6$

The critical values of this two-tailed test are ± 1.645. The test statistic is -1.333. $\alpha = .10$, while the p-value = .1855. The decision is to not reject H_0. At the 10% significance level, we cannot conclude that the mean time of employment is different from 6 months.

Note: This solution assumes you used a ***t*** distribution because sigma was unknown.

[For **TI-82**, use program **ONEPOPHT**. See page 100.]

Confidence Interval for the Difference of Means Using Summary Statistics and Normal Distribution

Example: A policeman is sitting on the side of the road with his radar gun and collecting data on speeds of cars and colors of cars. He has an idea that people who own red cars drive faster than people who own cars of other colors. He finds that the 90 red cars he has observed have a mean speed of 73.5 mph with a standard deviation of 9.87 mph. He finds that the 400 non-red cars he has observed have a mean speed of 68.9 mph with a standard deviation of 12.4 mph. Find the 95% confidence interval for the difference of the means of the speeds of red and non-red cars.

NOTE: With these large samples, the confidence interval will be constructed using a normal distribution where we approximate σ by **s**. Compare with the example on the next page.

FIRST: Organize your data, noting the summary statistics.

Population 1: People who drive red cars	Population 2: People who drive non-red cars
$\bar{x}_1 = 73.5$	$\bar{x}_2 = 68.9$
$\sigma_1 \approx s_1 = 9.87$	$\sigma_2 \approx s_2 = 12.4$
$n_1 = 90$	$n_2 = 400$

SECOND: Calculate the confidence interval from these values.

[For **TI-82**, use program **TWOPOP**. See page 100.]

Steps

STAT	
▸▸	Select **TESTS**.
9	Select **2-SampZInt**.
Stats	Move cursor to highlight **Stats**.
ENTER	
Enter the values	Enter the values shown, writing over old entries. Press **ENTER** after each entry. Approximate σ with **s**.

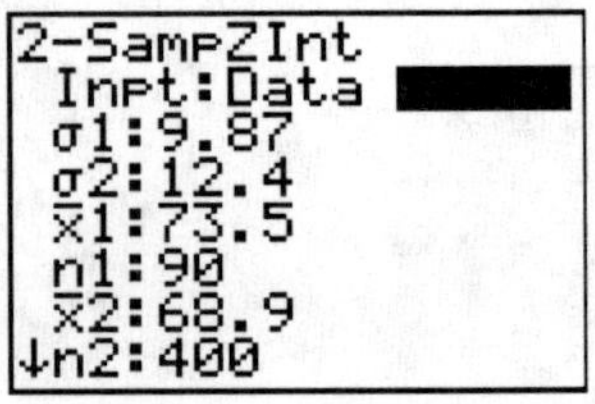

Steps	***(Continued)***	
▼	Scroll down and enter **C-Level**.	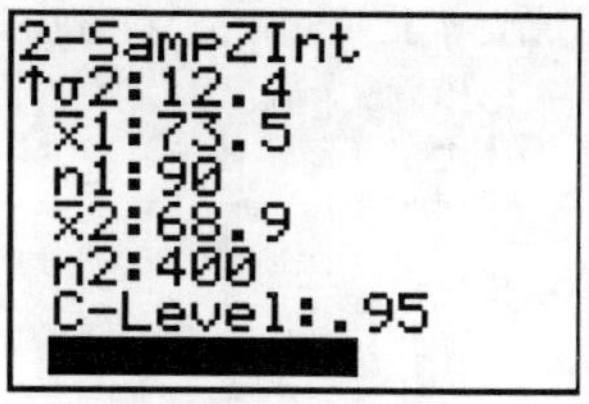
Calculate **ENTER**	Highlight **Calculate**.	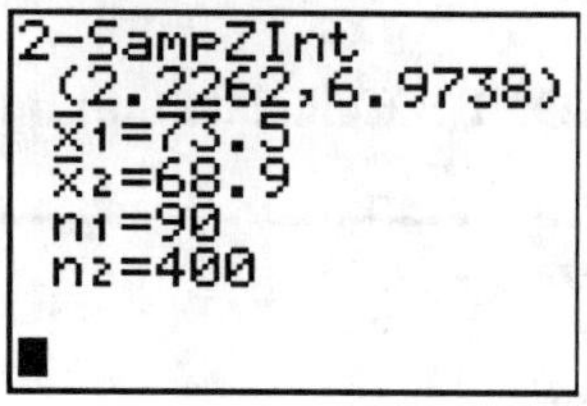

Solution: We are 95% confident that the difference in the mean speeds of red and non-red cars is between 2.2262 mph and 6.9738 mph.

Confidence Interval for the Difference of Means Using Summary Statistics and *t* Distribution

Example: A policeman is sitting on the side of the road with his radar gun and collecting data on speeds of cars and colors of cars. He has an idea that people who own red cars drive faster than people who own cars of other colors. He finds that the 90 red cars he has observed have a mean speed of 73.5 mph with a standard deviation of 9.87 mph. He finds that the 400 non-red cars he has observed have a mean speed of 68.9 mph with a standard deviation of 12.4 mph. Assuming that the population standard deviations are not equal, find the 95% confidence interval for the difference of the means of the speeds of red and non-red cars.

NOTE: Even though our samples are large, as we only have $\mathbf{s_1}$ and $\mathbf{s_2}$ (not σ_1 and σ_1), the confidence interval will be constructed using the $\boldsymbol{t}$ distribution. Compare with the example on the previous page.

[The **TI-82** can not use a ***t*** distribution for large samples.]

FIRST: Organize your data, noting the summary statistics.

Population 1: People who drive red cars

$\bar{x}_1 = 73.5$

$s_1 = 9.87$

$n_1 = 90$

Population 2: People who drive non-red cars

$\bar{x}_2 = 68.9$

$s_2 = 12.4$

$n_2 = 400$

SECOND: Calculate the confidence interval from these values.

Steps

STAT		
▶ ▶	Select **TESTS**.	
0	Select **2-SampTInt**.	
Stats	Move cursor to highlight **Stats**.	
ENTER		
Enter the values	Enter the values shown, writing over old entries. Press **ENTER** after each entry.	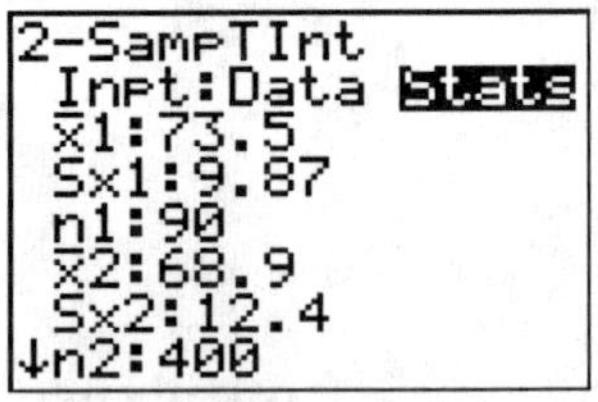
▼	Scroll down and enter **C-Level**.	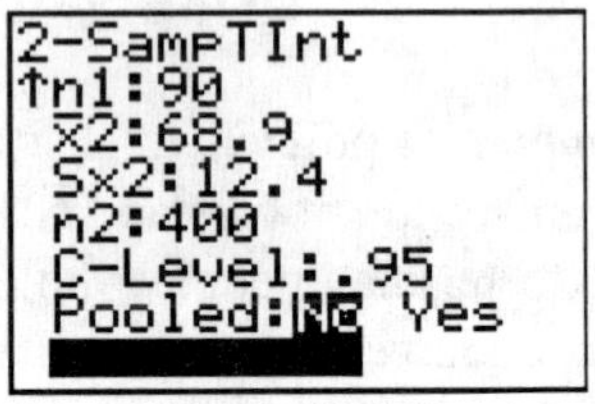
No **ENTER**	Highlight **POOLED:No** (because $\sigma_1 \neq \sigma_2$).	
▼ **Calculate** **ENTER**	Highlight **Calculate**.	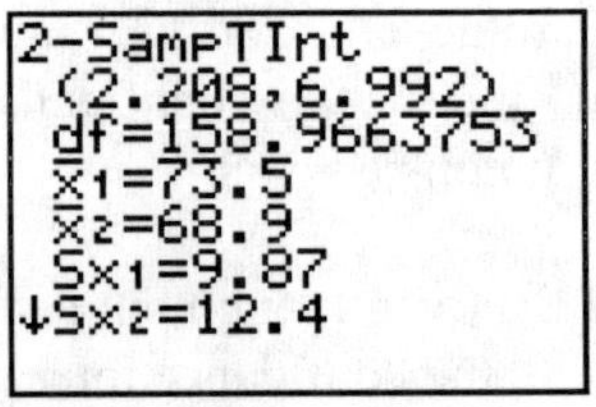

Solution: We are 95% confident that the difference in the mean speeds of red and non-red cars is between 2.208 mph and 6.992 mph.

Hypothesis Test about the Difference of Means Using Summary Statistics and *t* Distribution

Example: A policeman is sitting on the side of the road with his radar gun and collecting data on speeds of cars and colors of cars. He wants to test the hypothesis that people who own red cars drive faster than people who own cars of other colors. He finds that the 90 red cars he has observed have a mean speed of 73.5 mph with a standard deviation of 9.87 mph. He finds that the 400 non-red cars he has observed have a mean speed of 68.9 mph with a standard deviation of 12.4 mph. Assuming that the population standard deviations are not equal, test at the 5% significance level the hypothesis that the mean speed of red cars is greater than the mean speed of non-red cars.

NOTE: Even though our samples are large, as we only have $\mathbf{s_1}$ and $\mathbf{s_2}$ (not σ_1 and σ_1), this is a one-tail ***t*** test.

[The **TI-82** can not use a ***t*** distribution for large samples.]

FIRST: Set up the problem: $H_0: \mu_1 - \mu_2 = 0$ $H_1: \mu_1 - \mu_2 > 0$

SECOND: Organize your data, noting the summary statistics.

Population 1: People who drive red cars	Population 2: People who drive non-red cars
$\bar{x}_1 = 73.5$	$\bar{x}_2 = 68.9$
$\mathbf{s_1} = 9.87$	$\mathbf{s_2} = 12.4$
$\mathbf{n_1} = 90$	$\mathbf{n_2} = 400$

THIRD: Calculate the p-value and test statistic.

Steps

STAT	
▸▸	Select **TESTS**.
4	Select **2-SampTTest...**.
Stats	Move cursor to highlight **Stats**.
ENTER	
Enter the values	Enter the values shown, writing over old entries. Press **ENTER** after each entry.

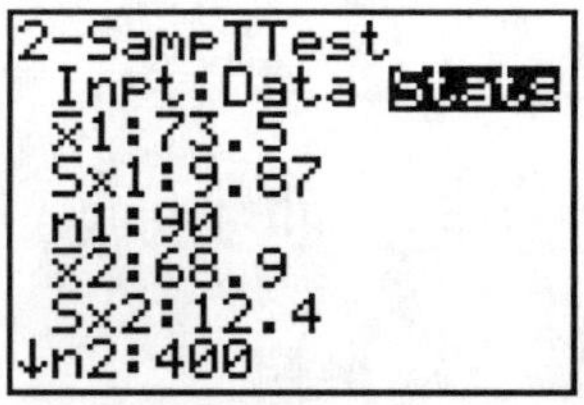

Steps	***(Continued)***	
▼ $>\mu_2$ **ENTER**	Scroll down and highlight the type of alternative hypothesis: $>\mu_2$.	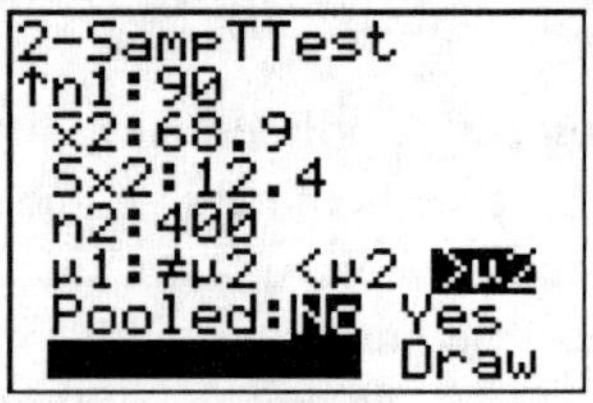
▼ **No** **ENTER**	Highlight **POOLED:No** (because $\sigma_1 \neq \sigma_2$).	
▼ **Calculate** **ENTER**	Highlight **Calculate**.	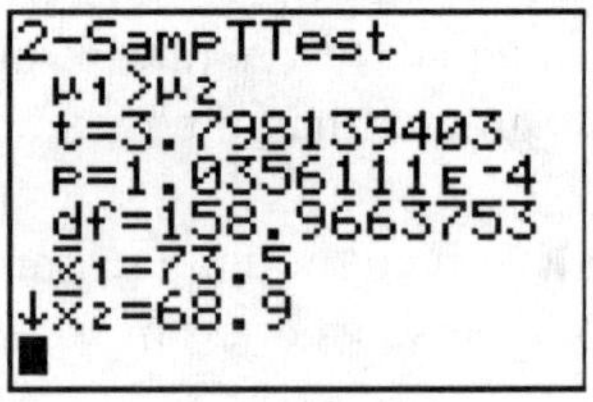

Decision: Reject H_0. The significance level = 5%, and the p-value = .00010356. The critical value = 1.655, and the test statistic = 3.798. There is sufficient evidence to conclude that the mean speed of red cars is greater than the mean speed of non-red cars.

Hypothesis Test about the Difference of Means Using Raw Data and *t* Distribution

Example: There is a widely held belief that female college students watch less sports on TV than do college males. To test this idea, a researcher asks 26 randomly chosen women and 19 men how many hours of TV sports they have watched in the last month. Their responses are below. Assuming that the population standard deviations are not equal, test at the 1% significance level the hypothesis that the mean TV time for women is less than the mean TV time for men.

Responses from Women: 3 5 4 7 6 12 9 2 5 3 8 10 15
7 9 7 12 10 6 8 2 16 0 9 11 7

Responses from Men: 7 10 11 15 16 17 23 1 6 2 16 12 8
4 7 19 11 3 14

FIRST: Set up the problem: Let Population 1 = women
& Population 2 = men

$H_0: \mu_1 - \mu_2 = 0 \qquad H_1: \mu_1 - \mu_2 < 0$

SECOND: Clear old data from List 1 and from List 2. Enter the data for women into List 1, and for men into list 2. Double check your data entry.

THIRD: Calculate the p-value and test statistic.

[For **TI-82**, use program **TWOPOP**. See page 100.]

Steps

STAT	
▶ ▶	Select **TESTS**.
4	Select **2-SampTTest...**.
Data	Move cursor to highlight **Data**.
ENTER	
Enter the values	Enter the values shown, writing over old entries. Press **ENTER** after each entry.
▼	Scroll down and highlight the type of
$<\mu_2$	alternative hypothesis: $<\mu_2$.
ENTER	
▼	
No	Highlight **POOLED:No**
ENTER	(because $\sigma_1 \neq \sigma_2$).
▼	Scroll down.
Calculate	Highlight **Calculate**.
ENTER	

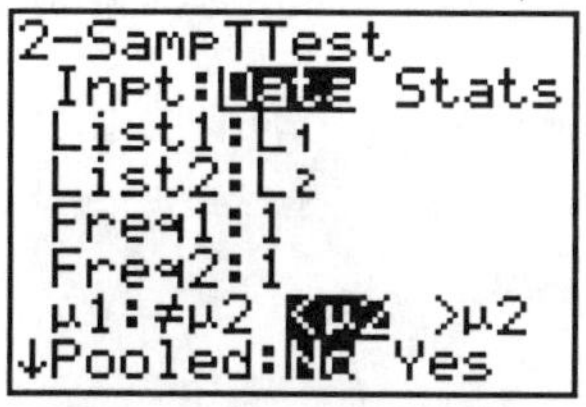

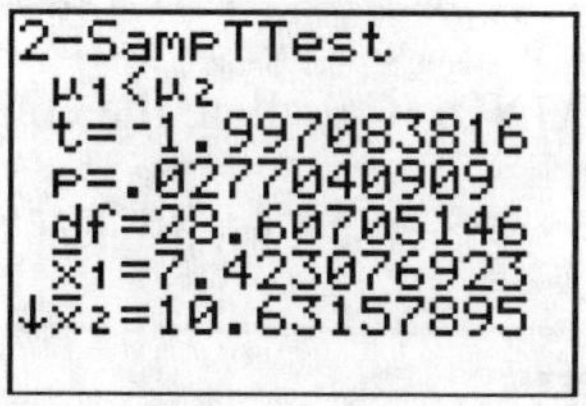

Decision: Do not reject H_0. The significance level = 1%, and the p-value = .0277. The critical value is t = -2.467, and the test statistic = -1.997. There is not sufficient evidence to conclude that the mean TV time for women is less than the mean TV time for men.

Confidence Interval for the Difference of Proportions

Example: In a recent study, American college students were asked if their social lives were satisfying. 800 Statistics students were asked, and 85% of them said yes. Of 1140 non-Statistics students asked, only 23% said yes. Construct the 95% confidence interval for the difference between the proportions of Statistics and non-Statistics students who report having satisfying social lives.

NOTE: As $\mathbf{n_1\hat{p}_1} > 5$, and $\mathbf{n_1\hat{q}_1} > 5$, and as $\mathbf{n_2\hat{p}_2} > 5$, and $\mathbf{n_2\hat{q}_2} > 5$, the samples are large and the normal distribution is used to construct the confidence interval.

FIRST: Organize your data, noting the summary statistics:

Population 1: Statistics students	Population 2: non-Statistics students
$\mathbf{x_1} = .85*800 = 680$	$\mathbf{x_2} = .23*1140 = 262.2 \approx 262$
$\mathbf{n_1} = 800$	$\mathbf{n_2} = 1140$
($\mathbf{\hat{p}_1} = .85$, and $\mathbf{\hat{q}_1} = .15$)	($\mathbf{\hat{p}_2} = .23$, and $\mathbf{\hat{q}_2} = .77$)

NOTE: Remember that $\mathbf{x = n\hat{p}}$. Furthermore, on the calculator, x must be an integer. If necessary, round x to the nearest integer.

SECOND: Calculate the confidence interval from these values.

[For **TI-82**, use program **TWOPOP**. See page 100.]

Steps

STAT		
▶▶	Select **TESTS**.	
ALPHA	Select **2-PropZInt....**	
B		
Enter the values	Enter the values shown, writing over old entries. Press **ENTER** after each entry.	2-PropZInt x1:680 n1:800 x2:262 n2:1140 C-Level:.95 Calculate
▼	Scroll down and enter **C-Level**.	

Steps	***(Continued)***	
Calculate ENTER	Highlight **Calculate**.	2-PropZInt (.58541,.65494) p̂1=.85 p̂2=.2298245614 n1=800 n2=1140

Solution: We can be 95% confident that the difference between the two proportions is between 58.5% and 65.5%.

Hypothesis Test about the Difference of Proportions

Example: In a recent study, American college students were asked if they were "fun loving." 780 Statistics students were asked, and 703 of them said yes. Of 1090 non-Statistics students asked, 805 said yes. At the 1% significance level, can we conclude that a larger proportion of Statistics students classify themselves as "fun loving," when compared to the proportion of non-Statistics students.

NOTE: As $\mathbf{n_1\hat{p}_1} > 5$, and $\mathbf{n_1\hat{q}_1} > 5$, and as $\mathbf{n_2\hat{p}_2} > 5$, and $\mathbf{n_2\hat{q}_2} > 5$, the samples are large and the normal distribution is used. The critical value of **z** is 2.326. The rejection region is to the right of this value.

FIRST: Set up the problem. $H_0: \mathbf{p_1 = p_2}$ $H_1: \mathbf{p_1 > p_2}$

SECOND: Organize your data, noting the summary statistics:

Population 1: Statistics students	Population 2: non-Statistics students
$\mathbf{x_1} = 703$	$\mathbf{x_2} = 805$
$\mathbf{n_1} = 780$	$\mathbf{n_2} = 1090$
($\mathbf{\hat{p}_1} \approx .90$, and $\mathbf{\hat{q}_1} \approx .10$)	($\mathbf{\hat{p}_2} \approx .74$, and $\mathbf{\hat{q}_2} \approx .26$)

THIRD: Calculate the p-value and test statistic.

[For **TI-82**, use program **TWOPOP**. See page 100.]

Steps

STAT		
▶ ▶	Select **TESTS**.	
6	Select **2-PropZTest....**	
Enter the values	Enter the values shown, writing over old entries. Press **ENTER** after each entry.	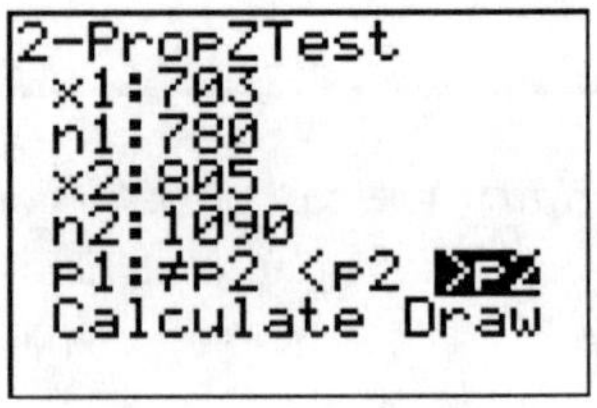
▼	Scroll down and highlight the type of	
$> p_2$ **ENTER**	alternative hypothesis: $> p_2$.	
Calculate **ENTER**	Move cursor to highlight **Calculate**.	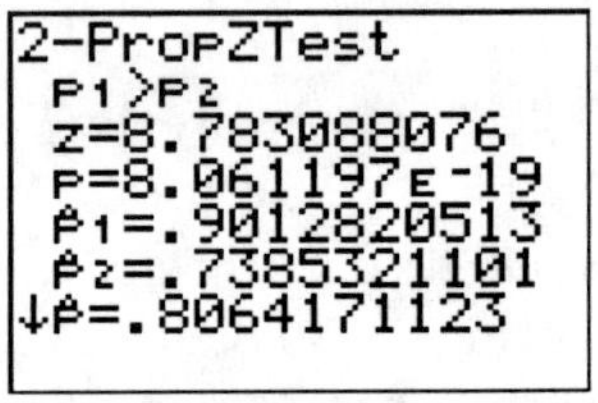

Decision: Reject H_0. The significance level = .01, and the p-value = .0000 (rounded). The critical value = 2.326, and the test statistic = 8.783. There is sufficient evidence to conclude that that the proportion of Statistics students who classify themselves as "fun loving" is greater than the proportion of non-Statistics students.

Hypothesis Test with Dependent Samples (Paired Differences) and *t* Distribution

Example: A track coach recorded 10 students' times in the mile before and after 6 weeks of training. Their times (in minutes) are given.

Before	13.83	15.83	12.05	8.19	9.78	9.82	9.92	14.21	12.12	15.27
After	10.43	7.48	6.61	8.99	5.85	4.96	6.63	7.65	12.13	11.13

Test at the 5% significance level if the training improved the running times.

FIRST: Set up the problem. $H_0: \mu_d = 0$ $H_1: \mu_d > 0$

SECOND: Enter the data into the lists. Clear List 1, List 2 and List 3. Put "Before" data in List 1 and "After" data in List 2. Put the ***paired differences*** in List 3.

Steps

L3 Move the cursor onto **L3** as shown.

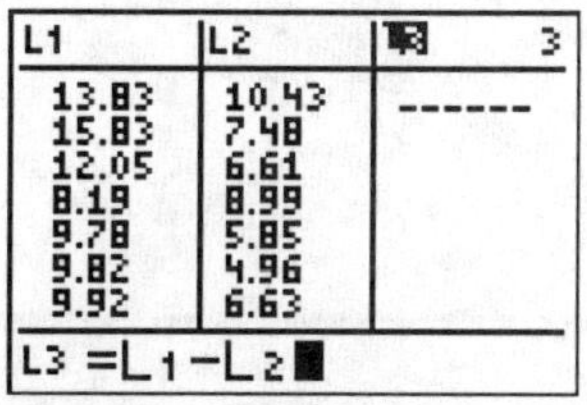

L1-L2

ENTER List 3 becomes List 1 - List 2, the paired differences.

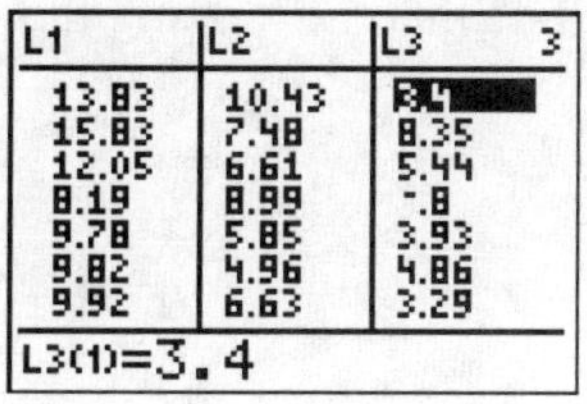

[For **TI-82**, use program **ONEPOPHT**. See page 100. For this **TI-82** program only, the paired differences <u>must</u> be in List 1. Follow the instructions above, then highlight **L1** at the top of List 1 and type **L3**. This will copy List 3 to List 1.]

THIRD: Calculate the p-value and test statistic.

Steps

STAT	
▶▶	Select **TESTS**.
2	Select **T-Test....**
Data	Move cursor to highlight **Data**.
ENTER	
Enter the values	Enter the values shown, writing over old entries. Press **ENTER** after each entry.
▼	Scroll down and highlight the type of
>μ_0	alternative hypothesis: **>**μ_0.
ENTER	
▼	Scroll down.
Calculate	Highlight **Calculate**.
ENTER	

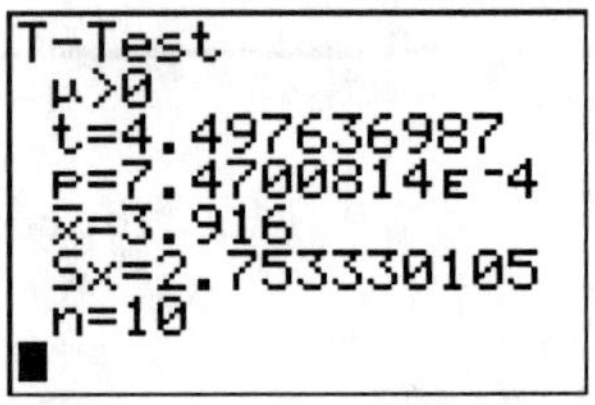

Decision: Reject H_0. The significance level = 5%, and the p-value = .000747. The critical value = 1.833, and the test statistic = 4.4976. There is sufficient evidence to conclude that the training did improve running times.

Supplementary Problems

1. In a study of college students, it was found that the mean GPA of 126 Republican students was 3.12 with a standard deviation of .78. The mean GPA of 340 non-Republican students was 2.98 with a standard deviation of 1.2. Find the 95% confidence interval for the difference in the mean GPAs of Republican and non-Republican students.

2. Ten students participated in a meditation class to see if it would lower their resting heart rate. Their resting heart rates before and after the meditation class are given in the following table.

Before	88	65	92	110	70	75	80	82	85	60
After	86	65	70	85	60	66	68	75	86	63

Find a 98% confidence interval for the mean of the paired differences of resting heart rates. Test at the 1% significance level, the claim that meditation improves (lowers) resting heart rates.

3. A student was doing a study on the cost of haircuts for males and females. Her survey had the following responses.

Males: $10 11 7.50 8.75 11 25 16 9.50 15 13.50 26 12.50

Females: $12 24 40 16 8.50 11.50 14.50 16 28 33

Assume that the populations are approximately normally distributed with equal standard deviations. Find a 99% confidence interval for the difference in the mean amount spent on haircuts for males and females. Test at the 5% significance level the hypothesis that females spend more on haircuts than males.

4. A study was conducted to see if retired people spend more time walking their dogs per day than non-retired people. The survey had the following responses (in hours).

Retired: 1.5 .75 .25 .5 1.25 3 4 1.75 2 .25 .5 1 .5 1.25 .5

Non-retired: .25 .5 1 2 2.5 .75 .5 .25 3 2.25 .25 .5 .25 .75 .5

Assume that the populations are approximate normally distributed with unequal standard deviations. Find a 95% confidence interval for the difference in the mean amount of time spent walking dogs per day by retired and non-retired people. Test at the 5% significance level the hypothesis that retired people spend more time walking their dogs per day than non-retired people.

5. A study of 1200 high school students found that 25% of them watched more than three hours of television a day. A study of 800 college students found that 16% of them watched more than three hours of television a day. Construct a 99% confidence interval for the difference in the proportions of high school students and college students who watch television more than three hours a day. Test at the 5% significance level the claim that the proportion of high school students who watch more than three hours of television a day is different from the proportion of college students who watch more than three hours of television a day.

Solutions for Supplementary Problems

CONFIDENCE INTERVAL, USING NORMAL DISTRIBUTION:

1a.

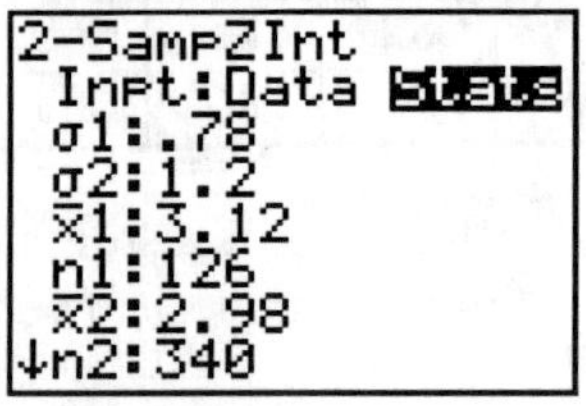

Enter summary statistics.

[For **TI-82**, use program **TWOPOP**. See page 100.]

Note: This solution assumes you used a normal distribution because the samples were <u>large</u>.

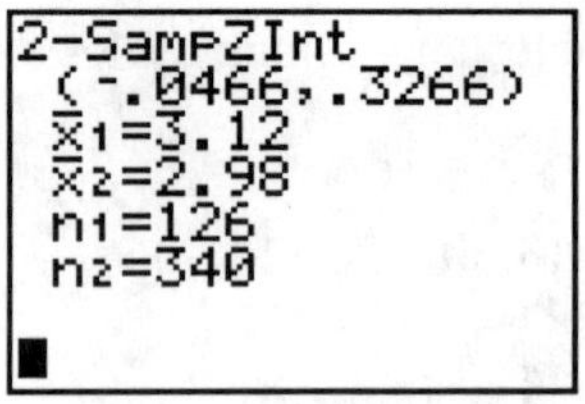

The 95% confidence interval for the difference in the mean GPAs of Republican and non-Republican students is -.0466 to .3266.

CONFIDENCE INTERVAL, USING ***t*** DISTRIBUTION:

1b.

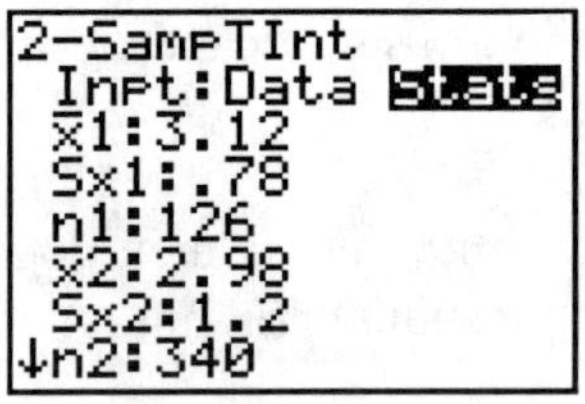

Enter summary statistics.
Do not use pooled σ's.

[The **TI-82** can not use a ***t*** distribution when **n** >30.]

Note: This solution assumes you used a ***t*** distribution because the sigmas were <u>unknown</u>.

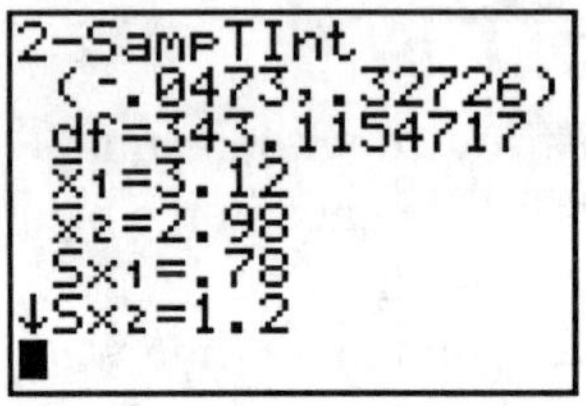

The 95% confidence interval for the difference in the mean GPAs of Republican and non-Republican students is -.0473 to .32726.

CONFIDENCE INTERVAL:

2.

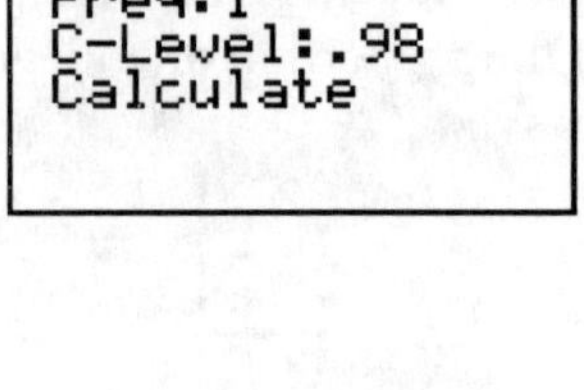

Input "Before" data to $\mathbf{L_1}$, and input "After" data to $\mathbf{L_2}$. Store *paired difference* data in $\mathbf{L_3}$ by using command $\mathbf{L_3} = \mathbf{L_2} - \mathbf{L_1}$. Then use **Tinterval**, and enter values.

[For **TI-82**, use program **ONEPOPCI**. See page 100.]

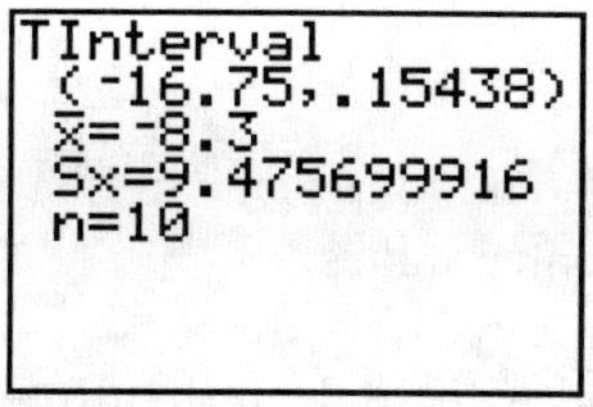

The 98% confidence interval for the mean of the paired differences of resting heart rates is -16.75 to .154.

TEST OF HYPOTHESIS:

2. 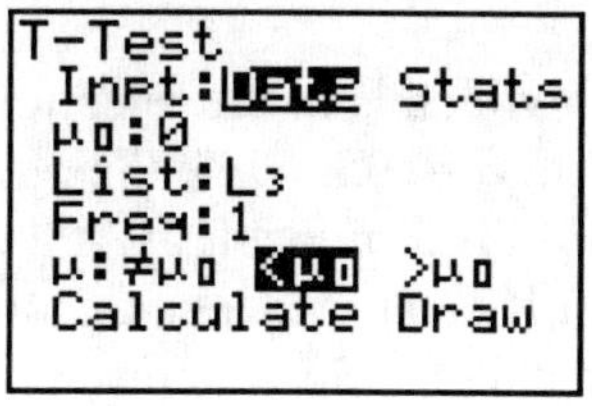

Calculate paired difference by using $L_3 = L_2 - L_1$.
Use **T-Test**, and enter values.
$H_0: \mu_d = 0$ $H_1: \mu_d < 0$
At the 1% significance level, with 9 degrees of freedom, the critical value is -2.821.

[For **TI-82**, use program **ONEPOPHT**. See page 100.
The paired differences must be in List 1.]

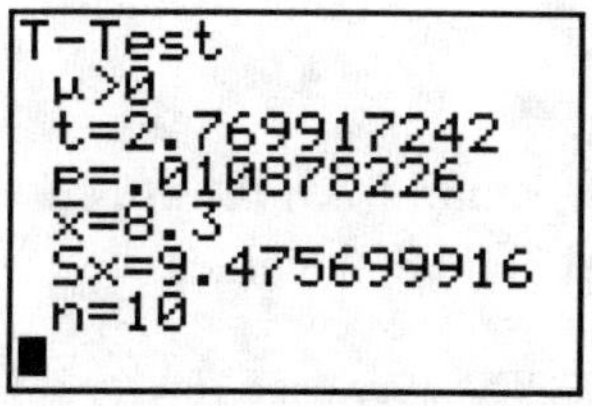

The test statistic is -2.7699 which is not in the rejection region for a left-tailed test. $\alpha = .01$, while the p-value = .0109. The decision is to not reject H_0. At the 1% significance level, there is not sufficient evidence to conclude that taking a meditation course lowers a person's resting heart rate.

CONFIDENCE INTERVAL:

3. 

Enter data for *men* in List 1, and for *women* in List 2.
Use **2-SampTInt** and enter values.

[For **TI-82**, use program **TWOPOP**. See page 100.]

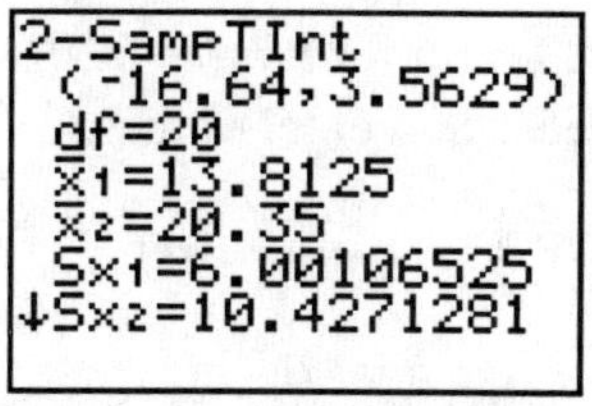

The 99% confidence interval for the difference in the mean amount spent on haircuts for males and females is -\$16.64 to \$3.56.

TEST OF HYPOTHESIS:

3. 

Enter data for *men* in List 1, and for *women* in List 2. Use **2-SampTTEST** and enter values.

$H_0: \mu_1 - \mu_2 = 0$ $H_1: \mu_1 - \mu_2 < 0$

At the 5% significance level with 20 degrees of freedom, the critical value is -1.725.

[For **TI-82**, use program **TWOPOP**. See page 100.]

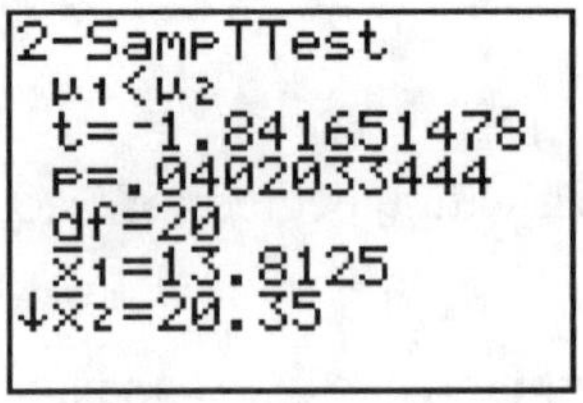

The test statistic is -1.84165 which is in the rejection region for this left-tailed test. $\alpha = .05$, while the p-value = .0402. The decision is to reject H_0. At the 5% significance level, there is sufficient evidence to conclude that men spend less on haircuts than do women.

CONFIDENCE INTERVAL:

4. 

Enter data for *retired* in List 1, and for *non-retired* in List 2. Use **2-SampTInt** and enter values.

[For **TI-82**, use program **TWOPOP**. See page 100.]

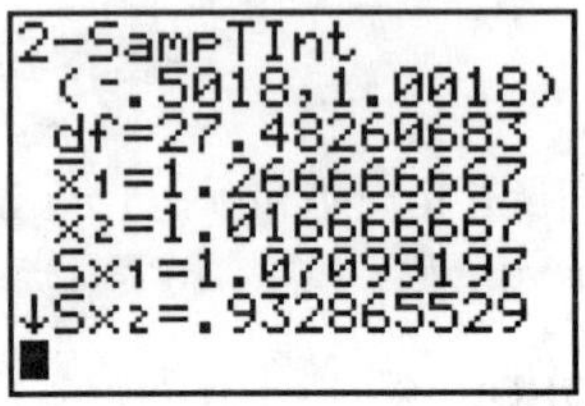

The 95% confidence interval for the difference in the mean time walking dogs per day by retired and non-retired people is from -.5018 to 1.0018 hours.

TEST OF HYPOTHESIS:

4. 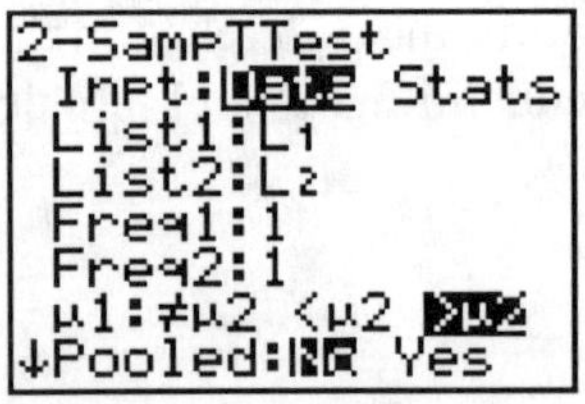

Enter data for *retired* in List 1, and for *non-retired* in List 2. Use **2-SampTTEST** and enter values.

[For **TI-82**, use program **TWOPOP**. See page 100.]

$H_0: \mu_1 - \mu_2 = 0$ $H_1: \mu_1 - \mu_2 > 0$

At the 5% significance level, with 27 degrees of freedom, the critical value is 1.703.

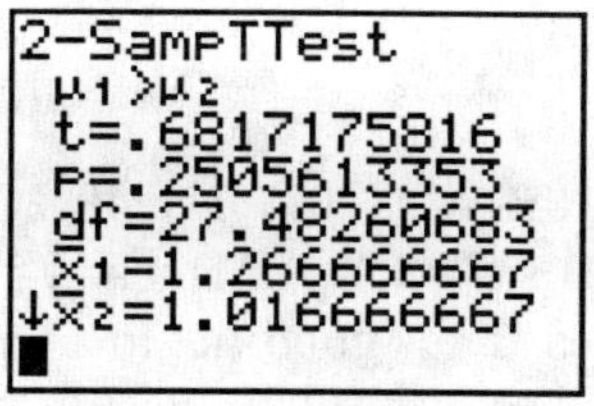

The test statistic is .6817 which is not in the rejection region. $\alpha = .10$, while the p-value = .2506. The decision is to not reject H_0. At the 5% significance level, there is not sufficient evidence to conclude that retired people spend more time per day walking their dogs than do non-retired people.

CONFIDENCE INTERVAL:

5.

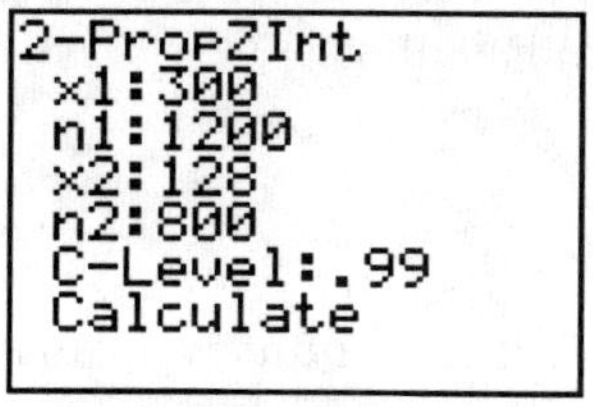

Use **2-PropZINT** and enter values.

$\mathbf{x_1}$ = 1200∗.25

$\mathbf{x_2}$ = 800∗.16

[For **TI-82**, use program **TWOPOP**. See page 100.]

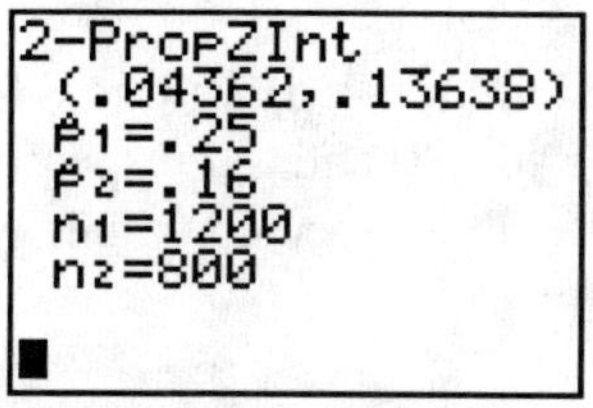

The 99% confidence interval for the difference in the proportions of high school students and college students who watch television more than three hours a day is between .04362 and .13638.

TEST OF HYPOTHESIS:

5.

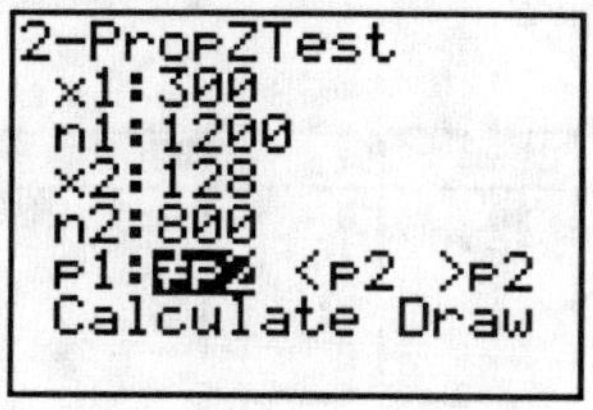

Use **2-PropZTest** and enter values.

[For **TI-82**, use program **TWOPOP**. See page 100.]

$H_0: \mathbf{p_1 - p_2} = 0$ $H_1: \mathbf{p_1 - p_2} \neq 0$

At the 5% significance level, the critical values are ± 1.960 for this two-tailed test.

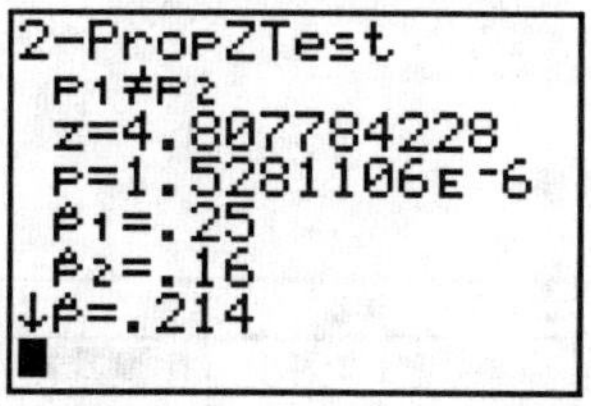

The test statistic is 4.8078 which is in the rejection region. $\alpha = .10$, while the p-value = .0000015. The decision is to reject H_0. At the 5% significance, there is sufficient evidence to conclude that the proportion of high school students who watch more than three hours of television a day is different from the proportion of college students who watch more than three hours of television a day.

Goodness-of-Fit Test Using Lists

***Example**:* Suppose that to grade on a curve means to give 10% A's, 20% B's, 40%C's, 20% D's and 10% F's. A teacher actually gave 10 A's, 18 B's, 35 C's, 12 D's and 5 F's. At the 5 % significance level would you say that these grades are a "good fit" to a curve.

FIRST: Set up problem: H_0: The grades are a good fit to a curve.
H_1: The grades are not a good fit to a curve.

A goodness-of-fit test uses the χ^2 distribution and is always right-tailed. This problem has 4 degrees of freedom. The critical value for this problem is 9.488.

The formula for the test statistic is $\sum \frac{(O-E)^2}{E}$.

SECOND: Clear old data from List 1, List 2 and List 3. Then enter *observed* values into List 1.

THIRD: Enter *expected* values into List 2. (recall: expected value = **np**) In this problem **n** = 80, and the probabilities are .10, .20, .40, .20 and .10 respectively, so all you need to do is type: 80∗.10 , 80∗.20 , etc. The expected values will display: 8, 16, 32, etc.

FOURTH: Calculate the test statistic:

Steps

L3	Highlight **L3** at the top of the column.
(**2nd** **L1** **-** **2nd** **L2** **)** **x²** **÷** **2nd** **L2** **ENTER**	Put **(O-E)²/E** in list 3.
2nd **QUIT**	Return to the home screen.

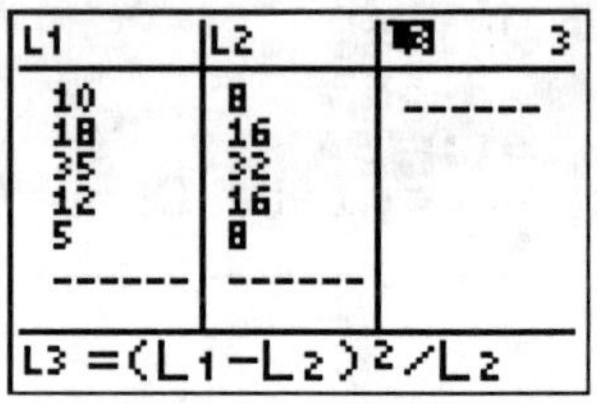

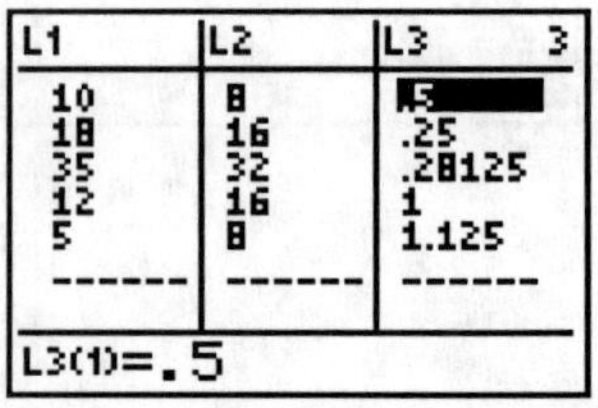

Steps	***(Continued)***
2nd	
LIST	
MATH	Move cursor to **MATH**.
5	Select **sum(**.
2nd	Select the sum of List 3.
L3	
)	[**)** is not required on **TI-82**.]
ENTER	
	The test statistic is 3.15625.

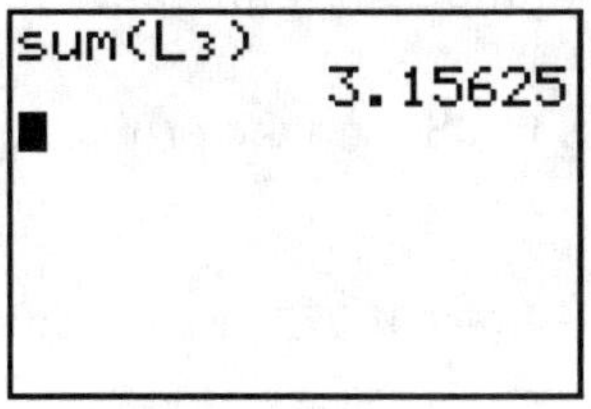

FIFTH: Calculate the p-value:

Steps	
2nd	Select the distribution menu.
DISTR	
7	Select χ^2**cdf(**. This function computes the cumulative probability from a lower bound to an upper bound.
3.15625	Enter the lower bound, the test statistic.
,	
10^99	Enter the upper bound.
,	(**10^99** approximates ∞.)
4	Enter the degrees of freedom.
)	
ENTER	

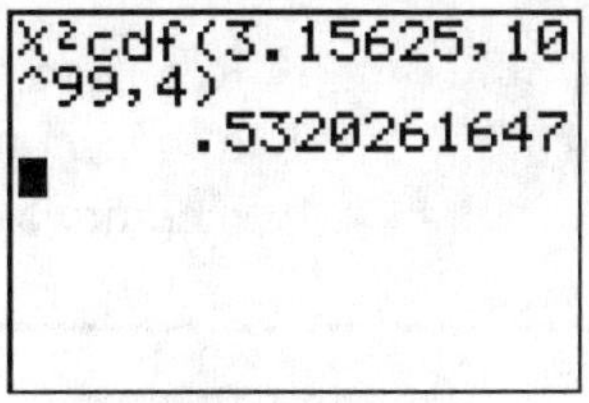

Decision: The test statistic is 3.15625. The critical value is 9.488 for this right-tailed test. The significance level is 5%, while the p-value is .532. The decision is to not reject H_0. The grades the teacher gave are a good fit to a curve.

(The program **GOODFT83** makes computing the test statistic easier. See page 98.)

[For **TI-82**, use the program **GOODFT82.** See page 98.]

Test for Independence

Example: Are the events gender and political party affiliation independent? A survey of 600 people yielded the following table of party affiliations.

	Republican	**Democrat**	**Other**
Male	150	125	20
Female	130	130	45

Test at the 5% significance level if gender and political party are independent.

FIRST: Set up the problem. H_0 : Gender and political party are independent.
H_1 : Gender and political party are not independent.

[For **TI-82**, use the program **CHISQ**. See page 99.]

SECOND: You use the **MATRX** editor to enter the *observed* matrix.

Steps

MATRX	
▶ ▶	Select **EDIT**.
1	Edit **MATRIX[A]** .
2	Specify the number of <u>rows</u>.
ENTER	
3	Specify the number of <u>columns</u>.
ENTER	
Enter the data	Enter data row by row. Press **ENTER** after each entry.
2nd **QUIT**	Return to the home screen. The matrix of *observed* values in now stored in memory as matrix **[A]**.

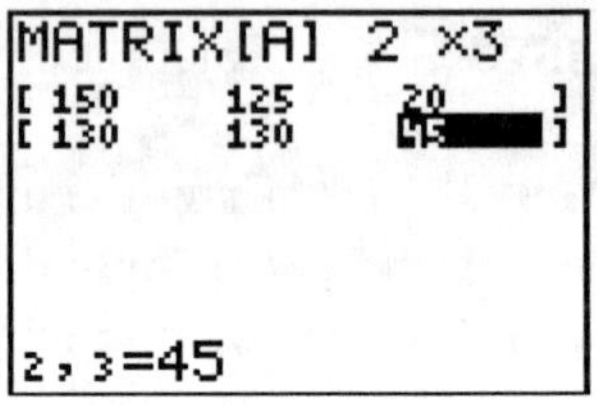

THIRD: Calculate the test statistic and the p-value.

Steps

STAT	
▸▸	Select **TESTS.**
ALPHA	Select χ^2 **-TEST....**
C	
Select Name of Observed Matrix	When **Observed:** matrix name is highlighted, use **MATRX**, ▲, ▼, and **ENTER** key to choose matrix **[A]**.
ENTER	
Select Name of Expected Matrix	When **Expected:** matrix name is highlighted, use **MATRX**, ▲, ▼, and **ENTER** key to choose matrix **[B]**. The *expected* matrix will be calculated automatically and it will be stored where you have just specified.
ENTER	
Calculate	Highlight **Calculate**.
ENTER	

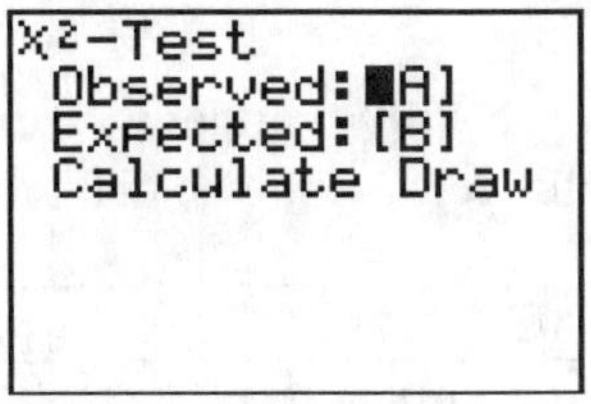

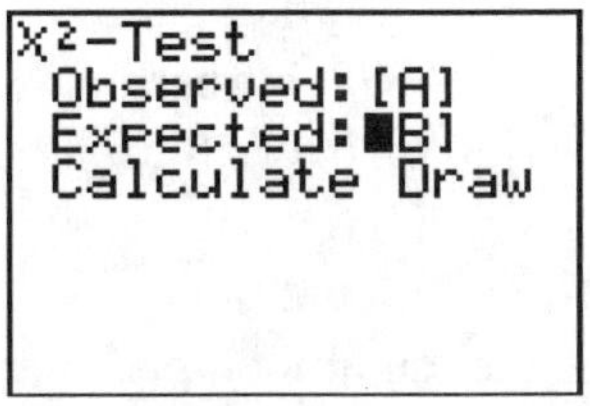

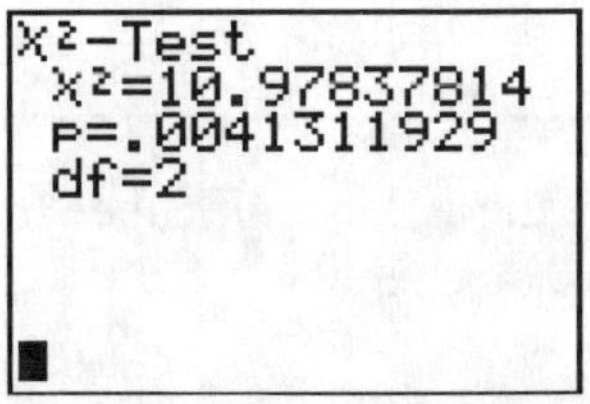

Decision: The critical value is 5.991, and the test statistic is 10.978. $\alpha = .05$, while the p-value = .0041. The decision is to reject H_0. Gender and political party are not independent.

Supplementary Problems

1. An anonymous diet claims that 40% of daily consumption should be fruits and vegetables, 15% should be meats, fish and poultry, 10% should be dairy products and 35% should be breads, cereals and pastas. Suppose that in one day, you consume 900 calories in fruits and vegetables, 500 calories in meats, fish and poultry, 200 calories in dairy products and 700 calories in breads, cereals and pastas. At the 2.5% significance level, is your consumption a good fit to the anonymous diet?

2. A cable TV company receives phone calls complaining about service. The following table represents the calls for one week.

Day of Week	**Sun.**	**Mon.**	**Tues.**	**Wed.**	**Thurs.**	**Fri.**	**Sat.**
No. of complaints	95	75	89	70	81	100	99

At the 5% significance level, test the null hypothesis that the orders are evenly distributed over all days of the week.

3. The following table is the result of a survey in which people were categorized by gender and blood pressure.

High Blood Pressure?	**Yes**	**No**
Male	240	300
Female	360	600

At the 10% significance level, can you conclude that gender and high blood pressure are independent?

4. The following table shows grades from two statistics classes.

Grades	**A**	**B**	**C**	**D**	**F**
Class 1	7	11	6	5	2
Class 2	10	12	7	4	2

Test at the 1% significance level if grades and class are related.

5. A survey was done in California, Nevada, Oregon and Arizona to ask motorists if they thought the speed limit should be raised.

State	**California**	**Nevada**	**Oregon**	**Arizona**
Raise speed limit	550	300	210	250
Do not raise limit	350	100	150	180

Using a 5% significance level, can you conclude that the distribution of opinions is homogeneous over these states?

Solutions for Supplementary Problems

1. 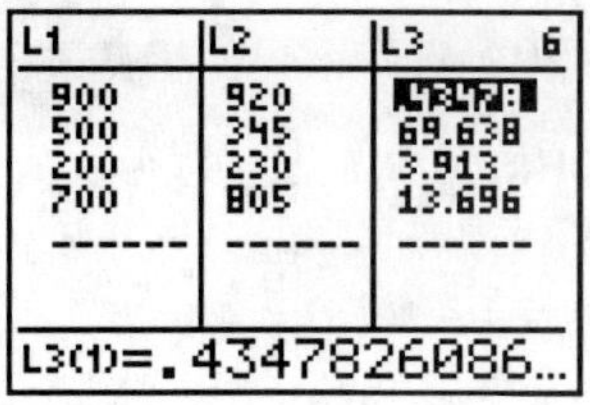

Enter observed values in List 1 and expected values in List 2. Define List 3 as **(L1 - L2)²/L2**.

H_0 : The daily consumption is a good fit to the anonymous diet.

H_1 : The daily consumption is not a good fit to the anonymous diet.

[For **TI-82**, use the program **GOODFT82**. See page 98.]

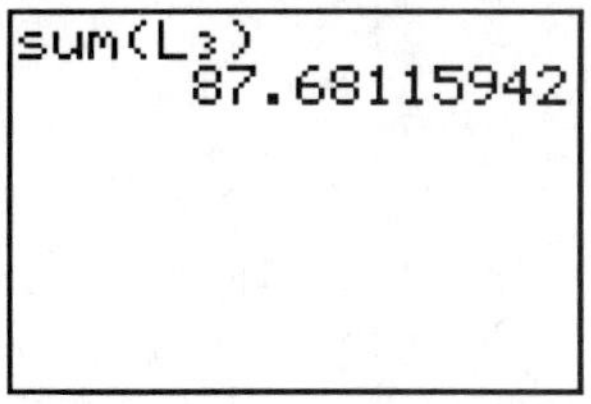

At the 2.5% significance level, with 3 degrees of freedom, the critical value is 9.348. The test statistic is 87.6812 which is in the rejection region.

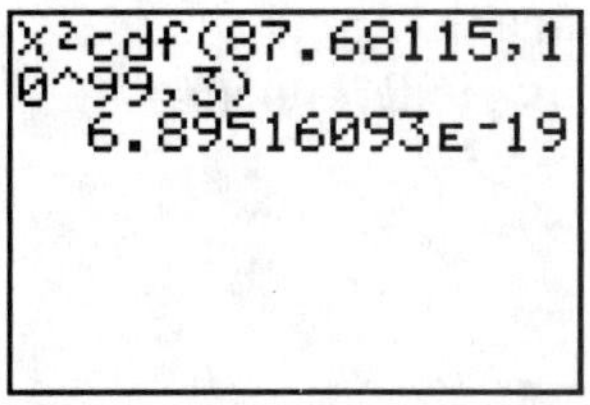

$\alpha = .025$, while the p-value = 0 (rounded). The decision is to reject H_0. At the 2.5% significance level, there is sufficient evidence to conclude that your daily consumption is not a good fit to the anonymous diet.

(Try the program **GOODFT83**. See page 98.)

2. 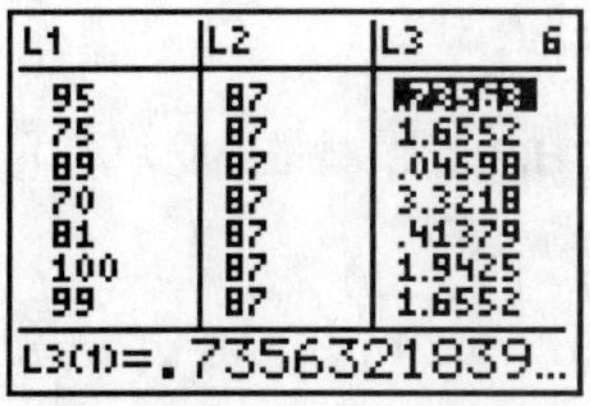

Enter observed values in List 1 and expected values in List 2. Define List 3 as **(L1 - L2)²/L2**.

H_0 : The complaints are evenly distributed throughout the week.

H_1 : The complaints are not evenly distributed throughout the week.

[For **TI-82**, use the program **GOODFT82**. See page 98.]

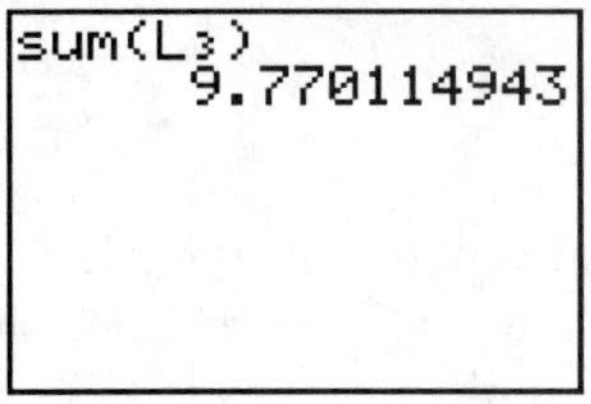

At the 5% significance level, with 6 degrees of freedom, the critical value is 12.592. The test statistic is 9.7701 which is not in the rejection region.

$\alpha = .05$, while the p-value = .1347. The decision is to not reject H_0. The complaints are evenly distributed throughout the week.

(Try the program **GOODFT83**. See page 98.)

3.

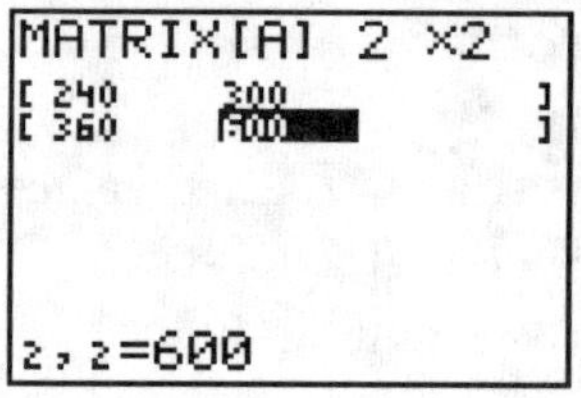

Enter *observed* values into matrix **[A]**.

H_0 : Gender & high blood pressure are independent.

H_1 : Gender & high blood pressure are not independent.

[For **TI-82**, use the program **CHISQ**. See page 99.]

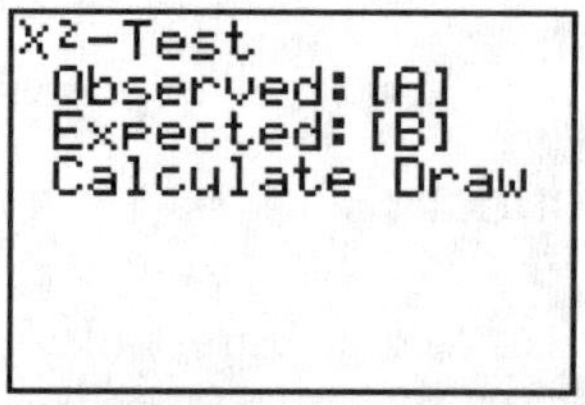

At the 10% significance level, with 1 degree of freedom, the critical value is 2.706.

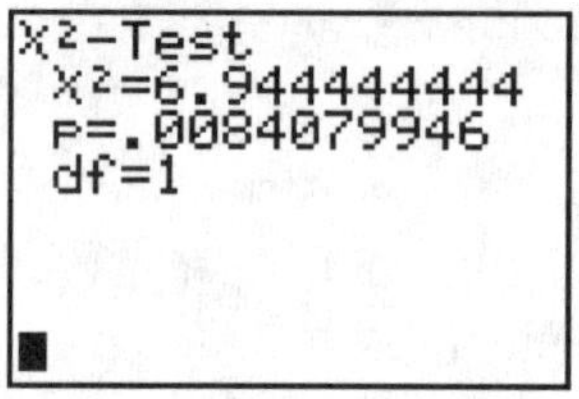

The test statistic is 6.9444 and is in the rejection region. $\alpha = .10$, and the p-value is .0084. The decision is to reject H_0. At the 10% significance level, there is sufficient evidence to conclude that gender and high blood pressure are not independent.

4.

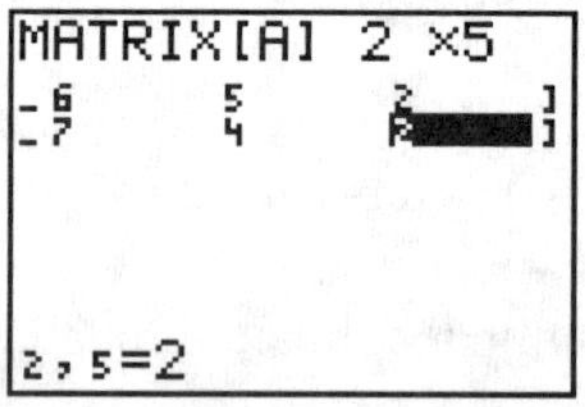

Enter *observed* values into matrix **[A]**.

H_0 : Grades and class are not related

H_1 : Grades and class are related.

[For **TI-82**, use the program **CHISQ**. See page 99.]

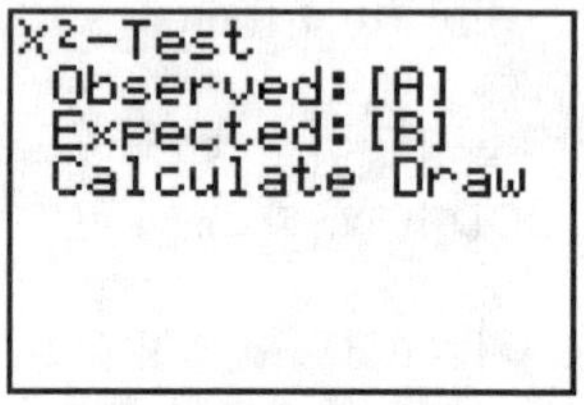

At the 1% significance level, with 4 degrees of freedom, the critical value is 13.277.

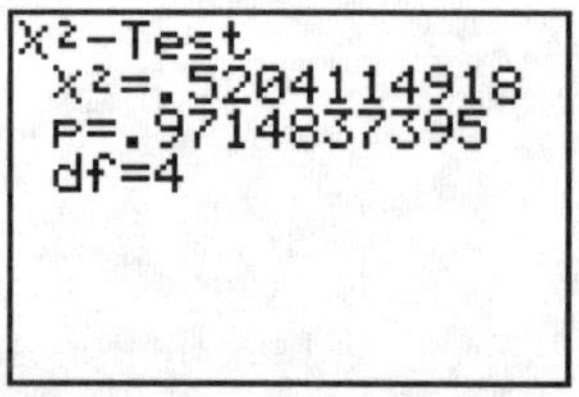

The test statistic is .5204 and is not in the rejection region. $\alpha = .01$, and the p-value is .9715. The decision is to not reject H_0. At the 1% significance level, there is not sufficient evidence to conclude that grades and class are related.

5. 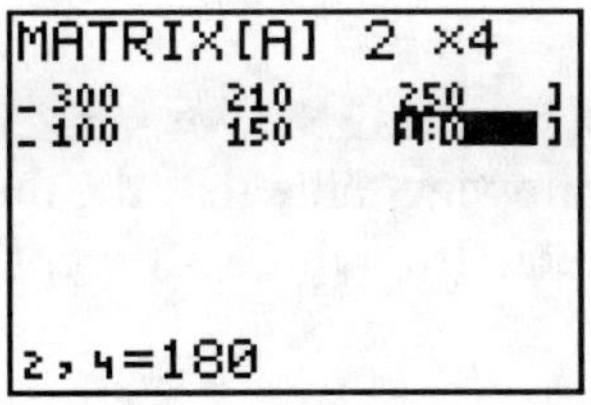

The test for homogeneity follows the same steps as the test for independence.

Enter *observed* values into matrix **[A]**.

H_0 : Distribution of opinions is homogeneous over these states.

H_1 : Distribution of opinions is not homogeneous over these states.

[For **TI-82**, use the program **CHISQ**. See page 99.]

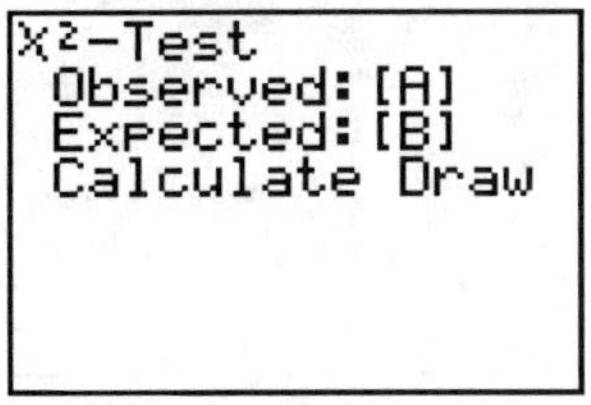

At the 5% significance level, with 3 degrees of freedom, the critical value is 7.815.

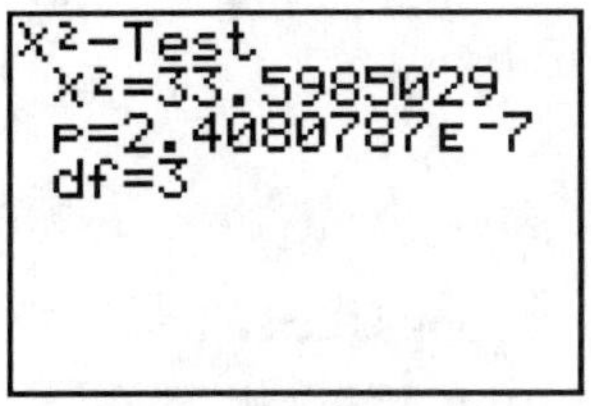

The test statistic is 33.5985 and is in the rejection region. $\alpha = .05$, and the p-value is .0000002. The decision is to reject H_0. At the 5% significance level, there is sufficient evidence to conclude that the distribution of opinions is not homogeneous over these states.

One-Way Analysis of Variance

Example: Girl Scout troops earn money by selling boxes of cookies. Sale organizers want to check if cookie sales per troop are the same. The following table lists the number of boxes of cookies sold by 7 randomly selected troop members from three different troops.

Troop 1	Troop 2	Troop 3
7	2	7
20	19	21
15	11	27
16	7	14
12	19	10
24	17	5
17	27	18

At the 5% significance level, can the organizers conclude that the mean number of cookie boxes sold by each troop is the same?

FIRST: Set up the problem.

H_0: The mean number of cookie boxes sold per troop is the same.
H_1: The mean number of cookie boxes sold per troop is not the same.

SECOND: Enter the data into List 1, List 2, and List 3.

THIRD: Compute the test statistic.

[For **TI-82**, use program **ANOVA**. See page 99.]

Steps

STAT	
►►	Select **TESTS**.
▲	Select **ANOVA(**.
ENTER	
2nd	Enter the Lists as shown.
L1	
,	
2nd	
L2	
,	
2nd	
L3	
)	

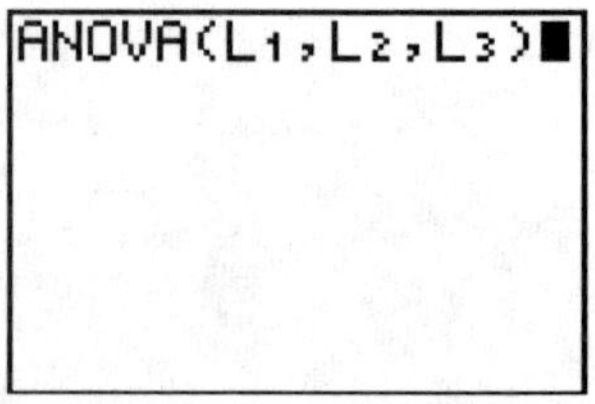

Steps	**(Continued)**	
ENTER	(The number of degrees of freedom of the numerator is 2.)	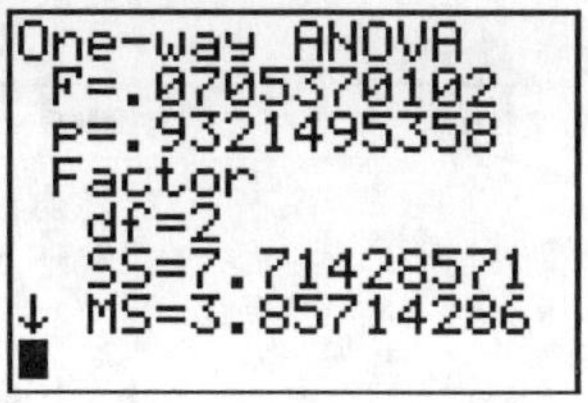
Scroll Down	Scroll down to the next screen. (The number of degrees of freedom of the denominator is 18.)	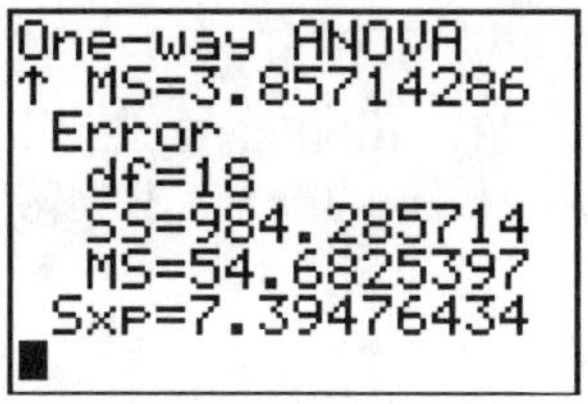

Solution: The critical value is 3.55, while the test statistic is .0705. The significance level is 5%, and the p-value is .9321. The organizers can continue to assume that the mean number of cookie boxes sold by each troop is the same.

Supplementary Problems

1. A college housing administer wants to know if the phone calls made in four different dorms are the same length. The following table shows the lengths (in minutes) of randomly selected phone calls from the four dorms.

Dorm 1	Dorm 2	Dorm 3	Dorm 4
24.4	14.1	6.5	21.2
27.3	18.6	15.4	17.4
30.2	25.7	14.6	23.9
22.6	7.4	8.7	20.6
33.7	16.1	3.3	8.3
21.9	23.9	9.9	17.5
23.1	24.2	11.0	13.2
9.0	10.3	14.6	
31.2		2.7	

At the 2.5% significance level, do the data indicate a difference in the means for the four dorms?

2. In an agricultural experiment 5 brands of plant food were used to enhance the growth of corn over a one week period. The following table gives the growth data in centimeters for randomly selected plants.

Brand 1	Brand 2	Brand 3	Brand 4	Brand 5
24.3	23.0	22.9	20.9	15.8
26.4	25.6	19.6	15.4	8.5
26.7	22.4	22.6	11.1	2.6
18.5	15.1	21.5	22.8	13.1
18.4	20.4	23.6	17.6	14.2
16.4	21.7	25.3	17.4	16.7

At a .01 significance level, is there a significant difference in the average growth of these plants for the different brands of plant food?

Solutions to Supplementary Problems

1.

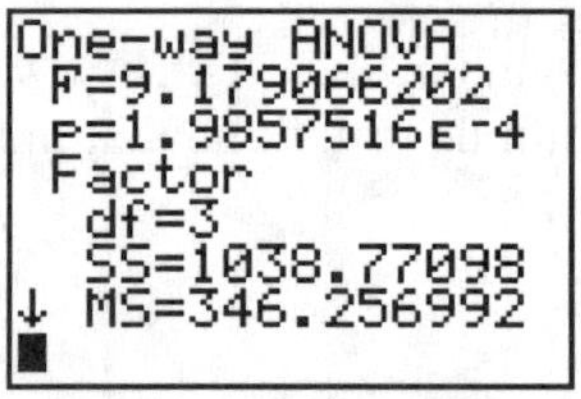

H_0: The means of the lengths of phone calls are the same.
H_1: The means of the lengths of phone calls are not the same.

[For **TI-82**, use program **ANOVA**. See page 99.]

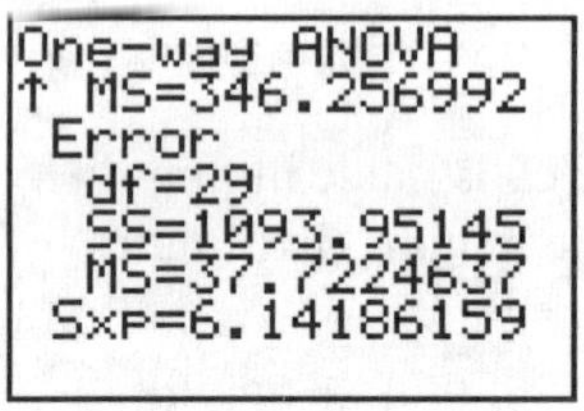

The degrees of freedom are (3,29). The critical value is 3.61, and the test statistic is 9.1791. The significance level is 2.5%, while the p-value is .0002. The decision is to reject H_0. The administrator may conclude that the mean phone call lengths are not the same for all four dorms.

2.

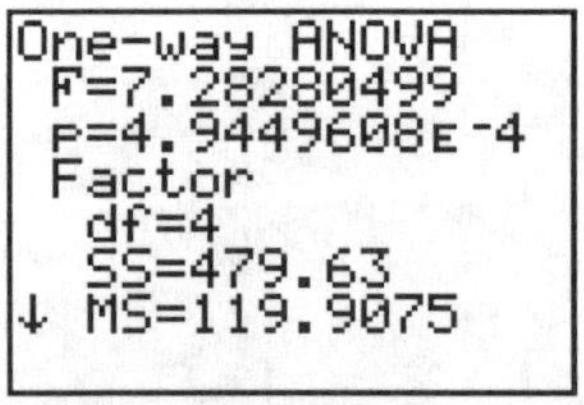

H_0: The means of the growth are the same.
H_1: The means of the growth are not the same.

[For **TI-82**, use program **ANOVA**. See page 99.]

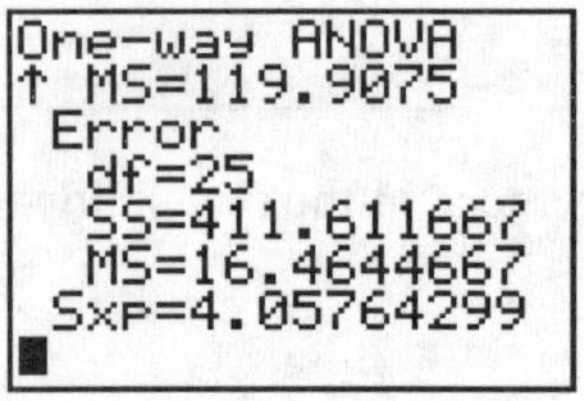

The degrees of freedom are (4,25). The critical value is 4.18, and the test statistic is 7.2828. The significance level is 1%, while the p-value is .0005. The decision is to reject H_0. We may conclude that there are significant differences in the growth rate means.

Linear Regression Equation and the Correlation Coefficient

Example: The following table shows the number of absences and mean test scores from 10 randomly selected students from an actual statistics class.

Absences	11	5	2	8	28	15	12	16	6	1
Mean Test Score	77.5	79.25	95.5	65.5	46.5	62.5	72	69.75	75	92.25

Find the regression equation, and the correlation coefficient, for this example.

FIRST: Clear old data from List 1 and List 2. Enter values of the *independent* variable (absences) in List 1. Then enter the values of the *dependent* variable (mean test scores) in List 2.

SECOND: Find **a**, the y-intercept, and **b**, the slope of the linear regression line.

Steps

STAT		
▶	Select **CALC** (for Calculations).	
8 [9 on TI-82]	Select **LinReg(a+bx)**.	
2nd **L1**	Let the calculator know that the **x**-values are in List 1.	
,		
2nd **L2**	Let the calculator know that the **y**-values are in List 2.	``LinReg(a+bx) L1, L2``
ENTER	Calculate **a** and **b**. [**TI-82** screen also shows **r**.]	LinReg y=a+bx a=90.23046335 b=-1.60148686

FOURTH: Find **r**. [This step is not necessary on the **TI-82**.]

Steps	
CLEAR	Clear the screen
VARS	Select memory area of **TI-83**.
5	Select **Statistics….**
▸ ▸	Select **EQ** (for Equations).
7	Select **r**.
ENTER	The value of **r** is retrieved.

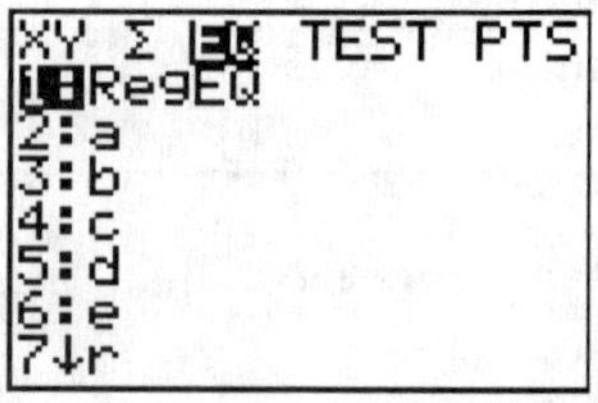

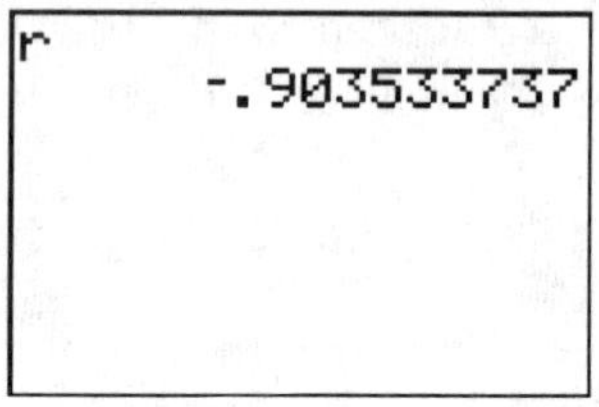

Solution: The equation of the regression line is: **y** = 90.23 - 1.60**x**. The correlation coefficient is -.9035, which is a strong negative correlation.

Automatic Display of the Correlation Coefficient

In the preceding example, the correlation coefficient **r** could have been displayed automatically when the intercept **a** and the slope **b** were calculated and displayed. On the **TI-83**, you need to set "**DiagnosticOn**." [This step is not necessary on the **TI-82**.]

Steps	
2nd CATALOG	Display all functions and variables available, some of which are <u>only</u> found in the **CATALOG**.
D	Skip forward through alphabet.
Scroll down	Scroll down until the pointer indicates **DiagnosticOn**.
ENTER ENTER	Select command **DiagnosticOn**.

Mann: 13.6

Graphing the Scatter Plot

FIRST: Turn off the graphs of any functions or **STAT PLOT**s currently turned on.

Steps

VARS	[On **TI-82**, use **2nd Y-VARS**.]
▶	Select **Y-VARS**.
4 [**5** on **TI-82**]	Select **On/Off…**.
2	Select **FnOff**.
ENTER	Turn off the graphing of previously stored functions.
2ND **STAT PLOT**	
4	Select **PlotsOff.**
ENTER	

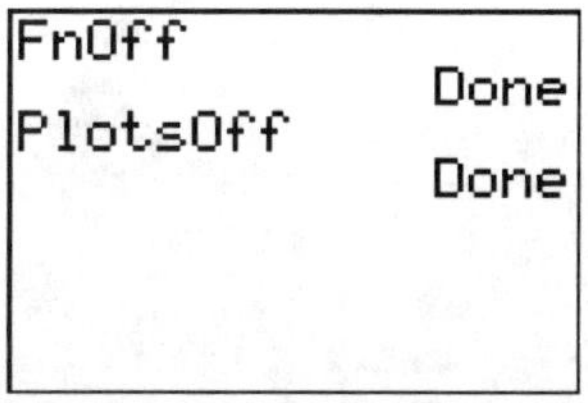

SECOND: Graph the scatter plot.

Steps

2nd **STAT PLOT**	**STAT PLOTS** controls the set-up for the scatter plot.
1	Select **Plot1** for configuration.
Use the cursor and ENTER	Set up your **Plot1** as shown. Make sure **On** is highlighted. [**TI-82** screens differ slightly.]
ZOOM **9**	Set the scale of the graph. Select **ZoomStat** and display the scatter plot on the screen.

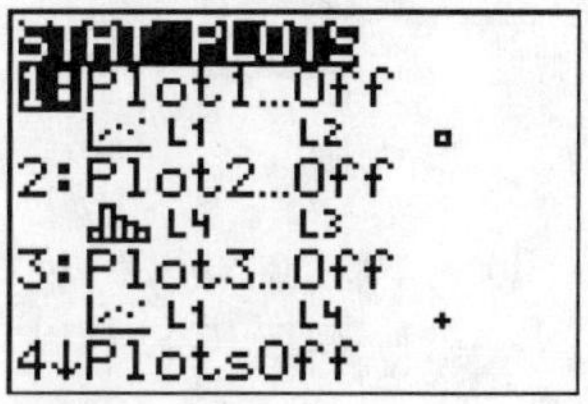

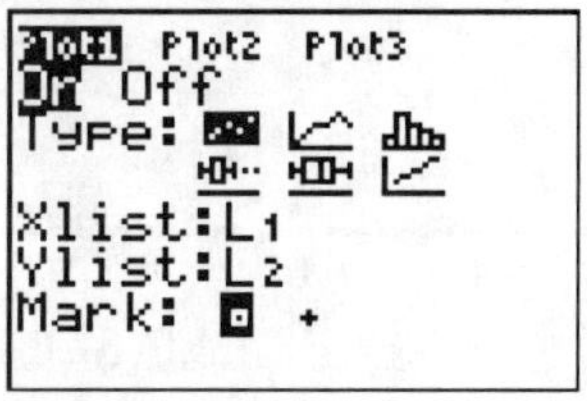

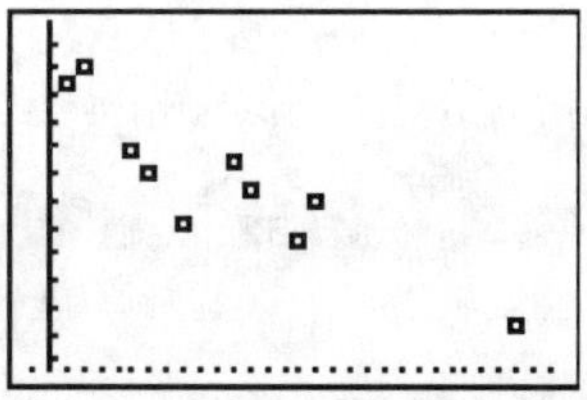

Graphing the Regression Line

Before graphing the regression line, you must have the *independent* variable data in List 1, the *dependent* variable data in List 2, and you must have computed **a,** and **b**.

Steps

Y=	Define the line you wish to graph. (Use the **CLEAR** key first, if needed.)	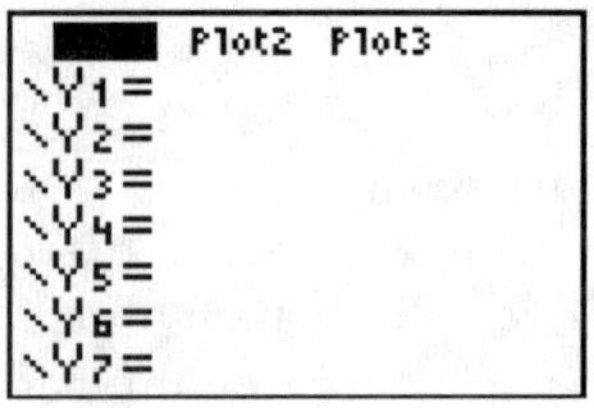
VARS **5** ▸ ▸	Access the variable's memory. Select **Statistics….** Select **EQ** (equation).	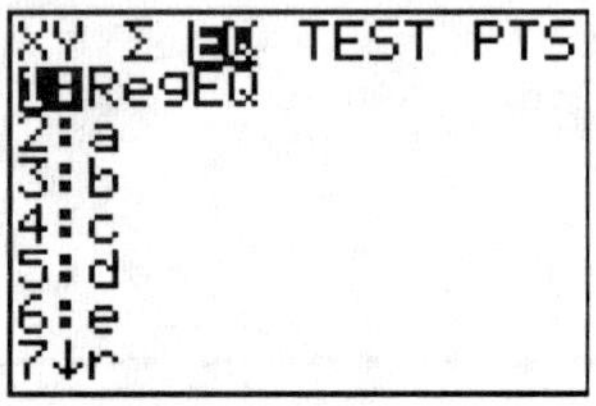
1 **[7 on TI-82]**	Select **RegEq** (regression equation) and paste it into **Y1**.	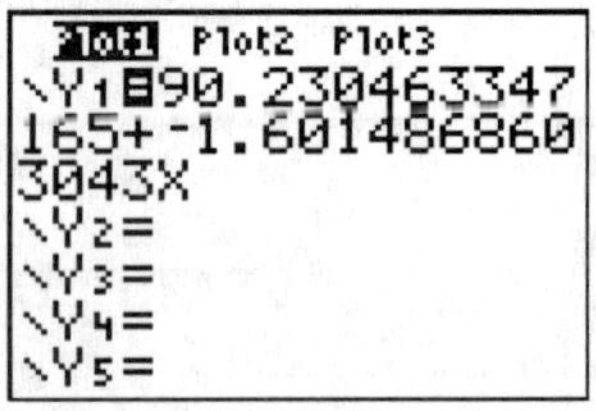
GRAPH	Display the scatter plot and the regression line on your screen.	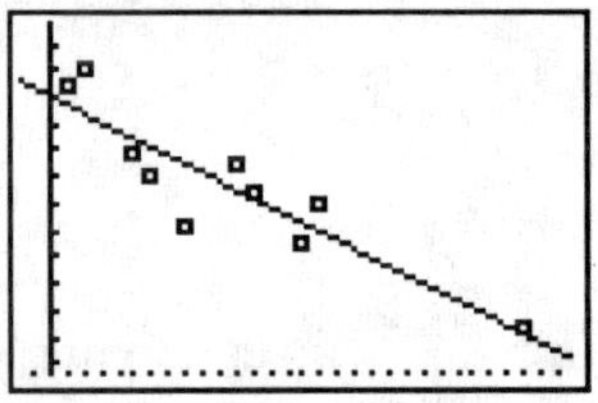

NOTE: In the example of this section, the command **LinReg(a+bx) L1, L2** was used to calculate the linear regression equation which was later retrieved from memory for graphing. As an alternative, you can calculate the regression equation, and store it in **Y1** simultaneously, by using the command **LinReg(a+bx) L1,L2,Y1**.

[On the **TI-82**, the **LinReg(a+bx)** command cannot automatically paste to **Y1**.]

Changing the Scale

Example: Show the **x**-axis from -10 to 40 by 5's and show the **y**-axis from -10 to 110 by 10's.

Steps

WINDOW	View current settings.	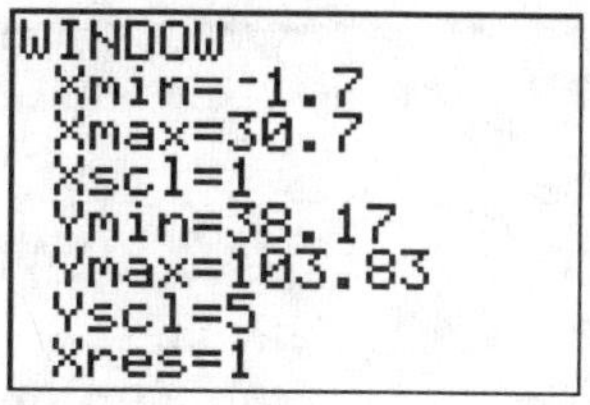
Enter new settings	Move the cursor, type in the new settings, and hit the **ENTER** key.	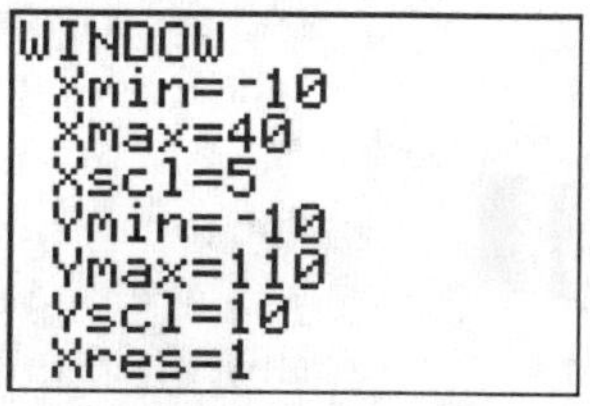
GRAPH	Display the scatter plot and the regression line in the <u>new</u> window.	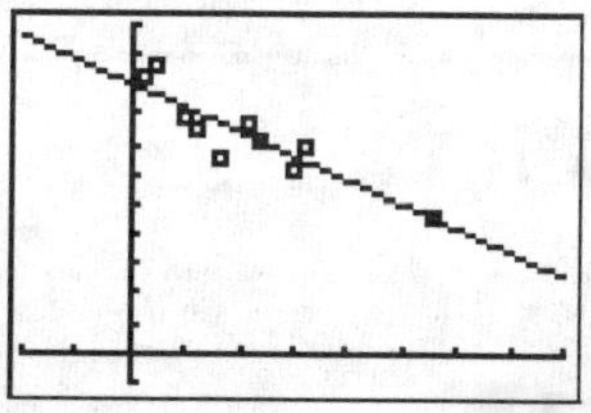

NOTE: You must choose your **WINDOW** settings carefully. For example, if you were to accidentally show the **x**-axis from -40 to 10 by 5's, and show the **y**-axis from -10 to 110 by 10's, your graph would appear different.

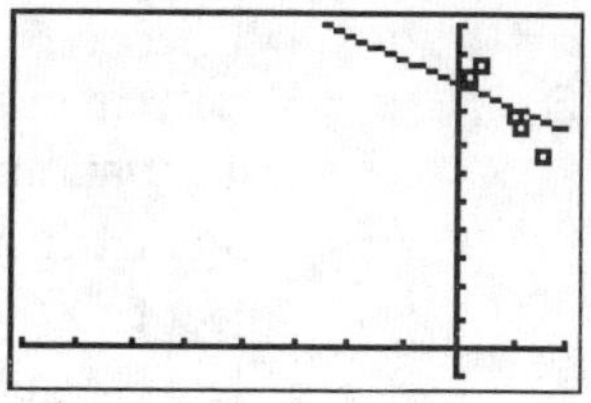

Prediction for a Given x-Value

Example: For the preceding example about absences and test averages, predict the mean test score for a student who is absent 20 times in a semester.

Method 1: (You need a graph of the regression line on the screen.)

Steps	
2nd	
CALC	**CALCULATE** on your just completed graph.
1	Select **value**.
20	Set the **x**-value = 20.
ENTER	The calculator displays the **y**-value at the bottom of the screen and the equation at the top. Therefore, we would predict a test average of 58.2 if a student were absent 20 times. [**TI-82** screens differ slightly.]

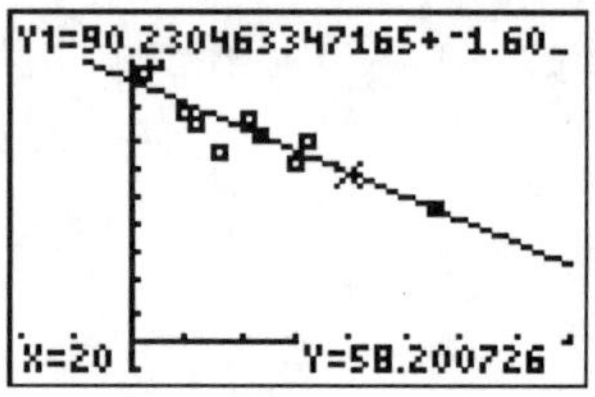

Method 2:

Steps	
CLEAR	Clear the screen.
20	
STO➡	Store 20 in **x**.
X,T,Θ,n	
ENTER	
VARS	View a menu of stored variable values and equations.
5	Select **Statistics…**.
▶ ▶	Select **EQ** (equation).
1 [**7** on **TI-82**]	Select **RegEQ** (regression equation).
ENTER	Evaluate the expression using the value of **x** just stored. Therefore, we would predict a test average of 58.2 if a student were absent 20 times.

Solution: If a student were absent 20 times, we would predict a test average of 58.2.

Mann: 13.2

Hypothesis Test About *B*

Example: Test at the 1% significance level if the slope of the regression line for the previous example relating average test scores and absences is negative.

Absences	11	5	2	8	28	15	12	16	6	1
Mean Test Score	77.5	79.25	95.5	65.5	46.5	62.5	72	69.75	75	92.25

[For **TI-82**, use program **HYPTESTB**. See page 99.]

FIRST: Set up the problem. $H_0 : B = 0$
$H_1 : B < 0$

SECOND: Enter the *independent* variables in List 1 and the *dependent* variables in List 2, if you have not already done so.

THIRD: Calculate the test statistic and the p-value.

Steps

STAT	
▶▶	Select **TESTS**.
▲▲	Select **LinRegTTest....**
ENTER	
Enter the values	Enter the values shown writing over old entries.
▼	Scroll down and highlight the type of
< 0	alternative hypothesis: **< 0**.
ENTER	
▼	Scroll down.
Calculate	Highlight **Calculate**.
ENTER	

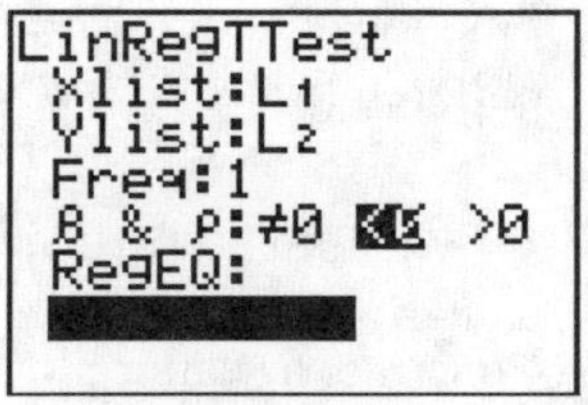

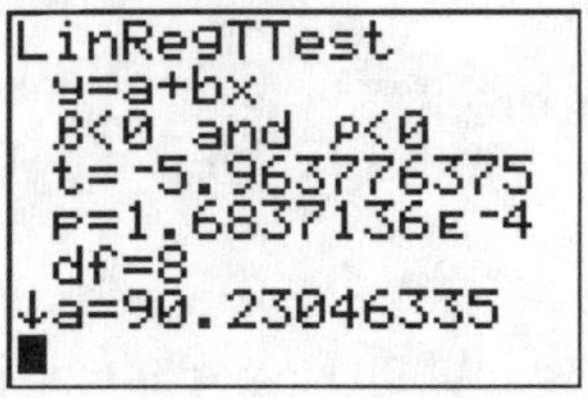

Solution: The critical value with 8 degrees of freedom is -2.896, while the test statistic is -5.9638. The significance level is 1%, while the p-value is .00017. The decision is to reject H_0. We conclude that test averages decrease as absences increase.

Supplementary Problems

A dozen families in a well-to-do city were asked:

1. How many children do you have?
2. How much money per month do you spend at the grocery store?
3. How many TV sets are in your home?
4. How many times per month does your family eat in a restaurant?

Their responses are summarized below:

children	1	4	2	3	2	7	1	2	5	2	6	2
groceries	450	1000	650	800	580	1350	550	590	1270	700	1380	720
TV sets	2	3	1	2	1	2	3	2	2	3	2	1
restaurants	4	2	6	2	4	1	5	3	2	3	1	2

1. a. Find the equation for the Linear Regression line (in the form **a + bx**) correlating children (*independent* variable) vs. groceries (*dependent* variable).

   b. Is the correlation positive or negative ? Is it strong or weak ?

   c. Construct a scatter plot of children vs. groceries and include the graph of the regression equation found in part a.

2. a. Find the equation for the Linear Regression line (in the form **a + bx**) correlating children (*independent* variable) vs. restaurants (*dependent* variable).

   b. Is the correlation positive or negative ? Is it strong or weak ?

   c. Construct a scatter plot of children vs. restaurants and include the graph of the regression equation found in part a.

   d. Using your regression equation as a predictor, how many times per month does a family of 3 eat in a restaurant? If $\mathbf{x} = 3$, $\mathbf{y} = ?$

3. a. Find the equation for the Linear Regression line (in the form **a + bx**) correlating children (*independent* variable) vs. TV sets (*dependent* variable).

   b. Is the correlation positive or negative ? Is it strong or weak ?

   c. Construct a scatter plot of children vs. TV sets and include the graph of the regression equation found in part a.

4. Test at the 5% significance level if the slope of the regression line for problem 3 relating number of children and number of TV sets is positive.

Solutions to Supplementary Problems

1a.
1b.

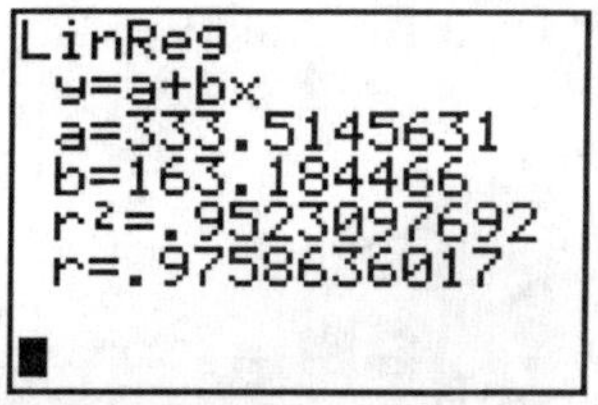

y = 333.515 + 163.184 **x**

r = .976 The correlation is positive and strong.

1c.

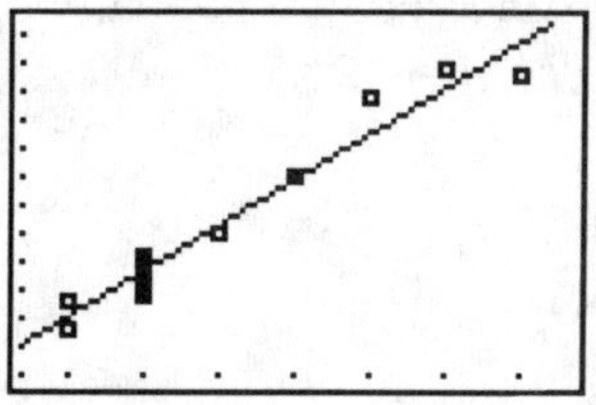

Graphs will vary with **WINDOW** settings.

The settings for this screen were set automatically by using the commands **ZOOM**, then **9**.

2a.
2b.

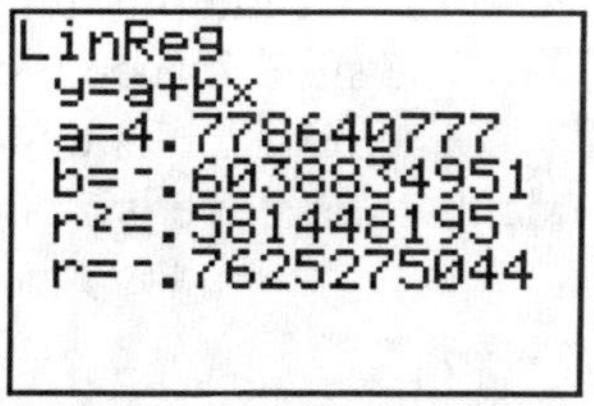

y = 4.779 - .604 **x**

r = -.7625 The correlation is negative and medium to strong.

2c.

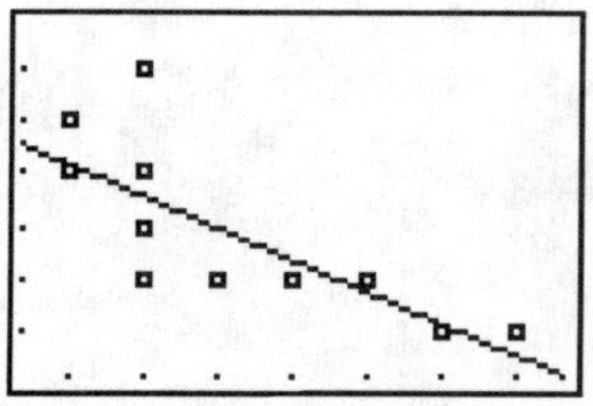

Graphs will vary with **WINDOW** settings.

The settings for this screen were set automatically by using the commands **ZOOM**, then **9**.

2d.

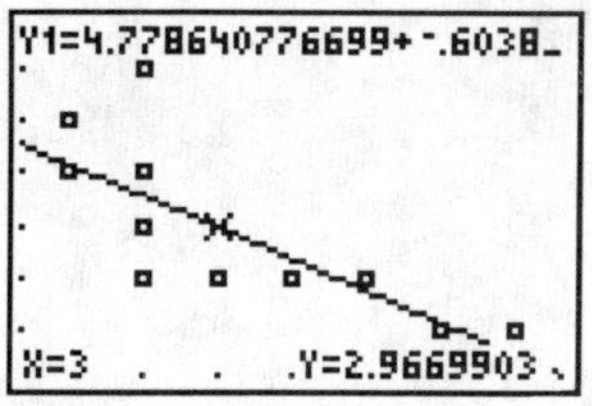

When **x** = 3, **y** = 2.967; so a family of 3 children is predicted to eat at a restaurant about 3 times per month.

3a.
3b.

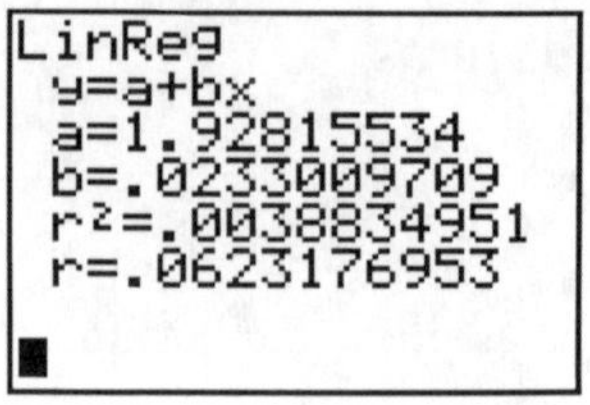

y = 1.928 + .0233 **x**

r = .0623 The correlation is positive and weak.

3c.

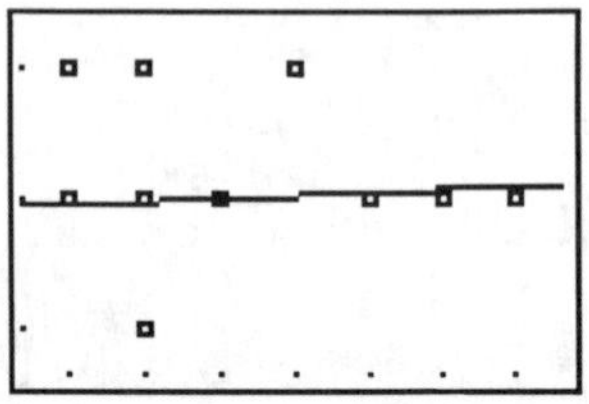

Graphs will vary with **WINDOW** settings.

The settings for this screen were set automatically by using the commands **ZOOM**, then **9**.

4.

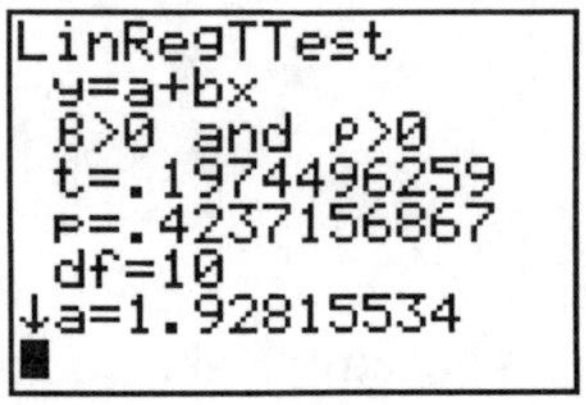

H_0 : $\boldsymbol{B} = 0$ H_1 : $\boldsymbol{B} > 0$

The critical value with 10 degrees of freedom is 1.812 while the test statistic is .1974. The significance level is 5%, while the p-value is .4237. The decision is to not reject H_0. We cannot conclude that the slope is positive.

How to Get the Programs

Programs can automate difficult calculations, reduce errors, and save you time. Five programs are modeled in this section [six programs for the **TI-82**] and are listed in the **Appendix**. These programs <u>must</u> be transferred to your calculator before you can use them. Any programs you receive from your instructor probably will be "write protected" so that they cannot be accidentally changed.

There are several methods to transfer the programs to your calculator. These methods are described below in order of ease.

1. The <u>easiest</u> way to transfer the programs to your calculator is to use the **LINK** command. To use the **LINK** command, you first need to have the programs on one calculator. It only takes a few seconds to transfer programs. The directions for using the **LINK** command are on the next page.

2. Instructors who have adopted this manual for their classes can receive a copy of the programs on disk by contacting John Wiley and Sons, Inc. The programs can be transferred from the computer disk to the calculator by using **TI-GRAPH LINK™** software by Texas Instruments which is available for free over the Internet at **http://www.ti.com**. In addition, you must have a special cable that connects your computer to the calculator. Call **1-800-TI-CARES** to get information on the cable. Once an instructor has the programs on one calculator, it is easy to transfer the programs to student calculators using the **LINK** command.

3. The calculator programs can be downloaded from the Internet. They are posted at three Web sites:

   http://www.wiley.com/college/mann W.A.V.E.
   http://home.earthlink.net/~gdeus
   http://home.earthlink.net/~johorner

   After you download the programs, you still will need the **GRAPH LINK™** software and the special cable described above.

4. The programs can be typed on your calculator from the listings in the Appendix. This method is quite slow and can lead to unintentional errors. The programs will not be "write protected" if they are put in the calculator using this method. See your **TI Guidebook©** to learn how to type in programs.

Program Basics

Pressing the [PRGM] key on your calculator will show a list of the programs stored in memory. They are in alphabetical order.

You start a program by pressing the number of the program, and then the [ENTER] key. After a program starts, you cannot go back and change entries you have made. So if you make an error, you have to stop and restart the program. The **QUIT** function does not always stop a program. While a program is running, press the [ON] key to stop the program. Then, this screen appears: Press either **1** or the [ENTER] key, and the program will stop. If you press the [ENTER] key again, then the program restarts.

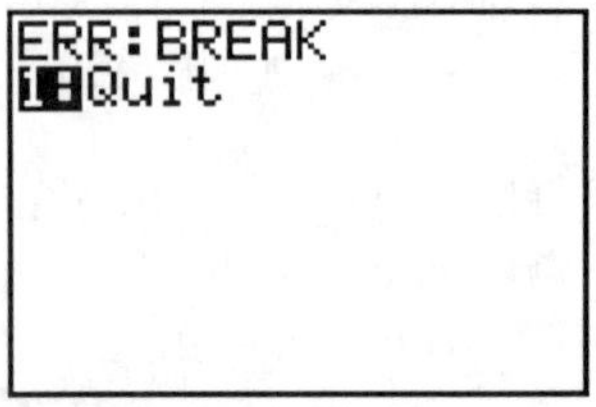

When a program is finished, the word **Done** appears on the screen. You may then go to another menu or do a calculation. When a program is **Done**, and you press the [ENTER] key, the program runs again. Press the [CLEAR] key to return to a blank home screen.

Using **Link** to Transfer Programs

Example*:** Transfer the five programs **BINOML83**, **GOODFT83**, **INVNOR83**, **NORMAL83**, **POISSN83**, from a ***sending calculator to a ***receiving*** calculator.

NOTE: You should only connect a **TI-83** with a **TI-83**, or a **TI-82** with a **TI-82**. The respective machines usually do not run programs for the other model without first making program modifications.

FIRST: Connect the two calculators with the cable. Be sure to press firmly to make a good connection.

SECOND: Set the ***receiving*** calculator to **Receive** the programs.

Steps	
2nd LINK	**LINK** the two calculators together.
▶	Highlight **RECEIVE**.
ENTER	Your calculator displays **Waiting…**.

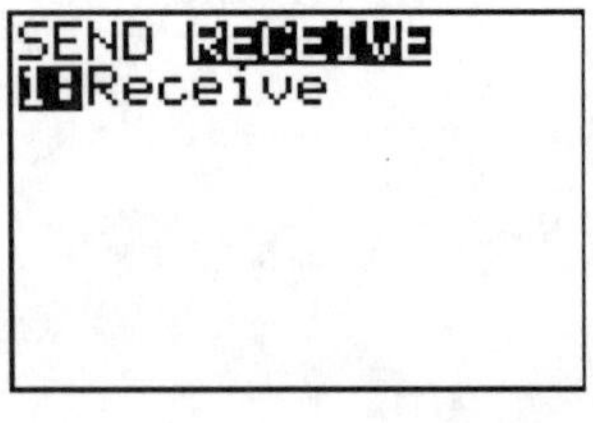

THIRD: Set the ***sending*** calculator to **Transmit** the programs.

Steps

2nd LINK — **LINK** the two calculators together.

3 **[2 on TI-82]** — Your calculator displays all the programs (**PRGM**) in the ***sending*** calculator. Screens may vary. You may choose **1** here and send <u>all</u> programs, lists and variables.

Select the Programs to Transmit — Move the cursor up and down until it is next to the programs you want to send. **ENTER** next to each and selection symbols (■) will mark your choices.

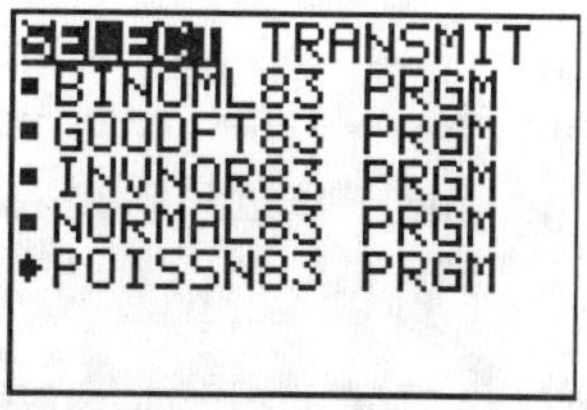

▸ ENTER — Highlight **TRANSMIT**. The programs you have just sent are listed on the screen.

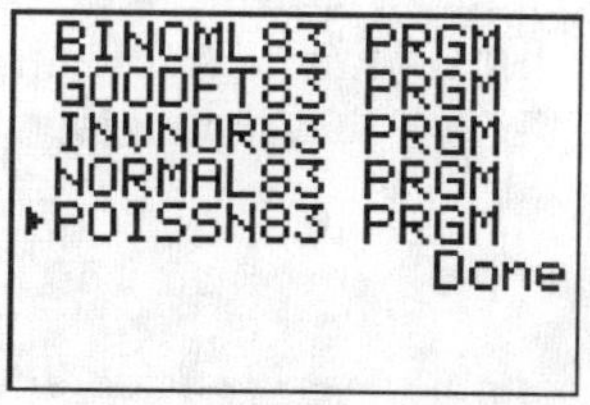

FOURTH: Check that the ***receiving*** calculator now has the programs.

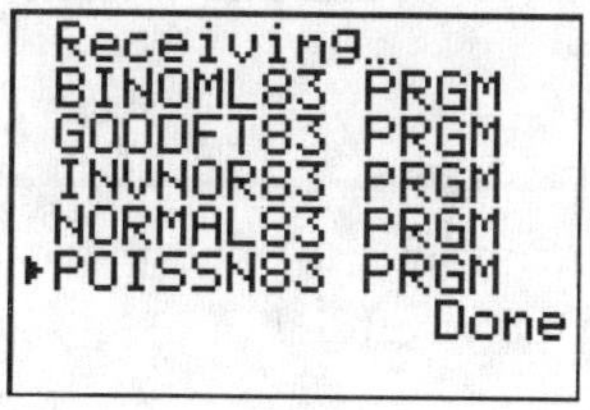

TROUBLESHOOTING:

If the ***receiving*** calculator already has a program with the same name as the one being sent, you must choose to **Rename** or **Overwrite** to complete the transfer. Use **Overwrite** to replace an older or corrupted version of the program you are receiving. Choose **Rename** to have both. If **Rename** is chosen, the common name is retained by the program already in memory, and you give a new name to the program being received.

If the calculator screens displays **Error in Xmit**, check that the cable is securely connected. Then start over.

How to Reset MODE and FORMAT

After using a program, especially if you quit early, your screen may look "**w*e*ird**," or something else may be "**ActiNg fu*nn*y**." If so, you should reset the **MODE** and **FORMAT**.

To reset the **MODE**, press the [MODE] key. Move the cursor to highlight the settings at the left side of the screen and hit the [ENTER] key to reset entries. When done your screen looks like the one shown here.

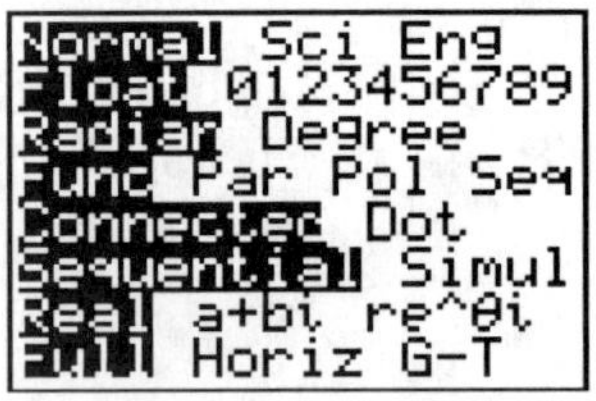

To reset the **FORMAT**, press **2nd FORMAT**. Move the cursor to highlight the settings at the left side of the screen and hit the [ENTER] key to reset entries. When done your screen looks like the one shown here.

If you continue to experience problems, you may need to check **STAT PLOT** and **WINDOW** settings.

Expected Value and Standard Deviation of a Discrete Probability Distribution Using the TI-82 Program PROBDIST

Example 1: Each day on a college campus, parking tickets are issued. Let **x** = the number of parking tickets issued in a day. The following table shows the probability distribution of **x**.

No. of tickets issued in a day	5	6	7	8	9	10	11	12
Probability	.12	.14	.10	.20	.15	.16	.08	.05

Find the ***expected number*** of parking tickets for a day. (The expected number, or expected value, is the same as the mean.) Also, find the standard deviation.

[**PROBDIST** is a program for the **TI-82** only. For **TI-83**, see page 15.]

Steps

PRGM	Select the program menu.	
PROBDIST	Enter the number of the program corresponding to **PROBDIST**.	
ENTER	Start the program **PROBDIST**.	
8	Enter the number of outcomes.	
Enter x values	Enter the **x** values. Press **ENTER** after each entry.	?5 ?6 ?7 ?8 ?9 ?10 ?11 ?12
Enter P(x) values	Enter the P(**x**) values. Press **ENTER** after each entry.	?.12 ?.14 ?.10 ?.20 ?.15 ?.16 ?.08 ?.05
	After the last data entry, the calculator computes and displays the mean and standard deviation of the discrete probability distribution.	MEAN = 8.17 ST DEV = 2.005268062 Done
CLEAR or **ENTER**	**CLEAR** to use another function, or **ENTER** to run **PROBDIST** again.	

Solution: The expected number of tickets (mean) is 8.17. The standard deviation is 2.01.

Mann: 5.3, 5.4

Binomial Probabilities Using the Program **BINOML83**

Example 1: Suppose a coin is tossed 10 times. Find the probability that "heads" come up exactly 6 times.

[**BINOML82** is the corresponding program for the **TI-82**.]

NOTE: For this problem the probability of success "**p**" is .5, the number of trials "**n**" is 10, and the number of successes "**x**" is 6.

Steps

PRGM **BINOML83**	Select the program menu. Enter the number of the program corresponding to **BINOML83**.
ENTER **1**	Start the program. Select **SINGLE P(X)**.

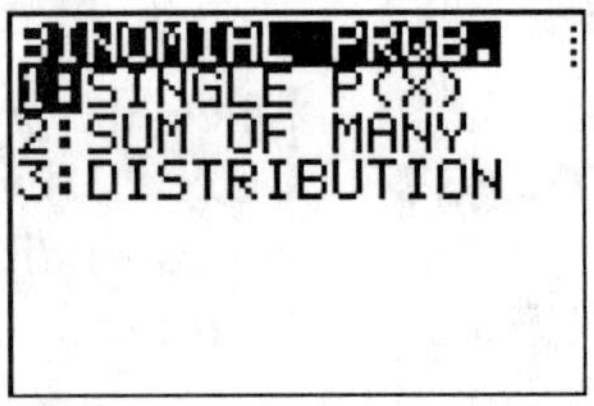

10 **ENTER** **.5** **ENTER** **6** **ENTER**	Enter the number of trials. (**n** = 10) Enter the probability of success on a single trial. (**p** = .5) Enter the number of successes. (**x** = 6)

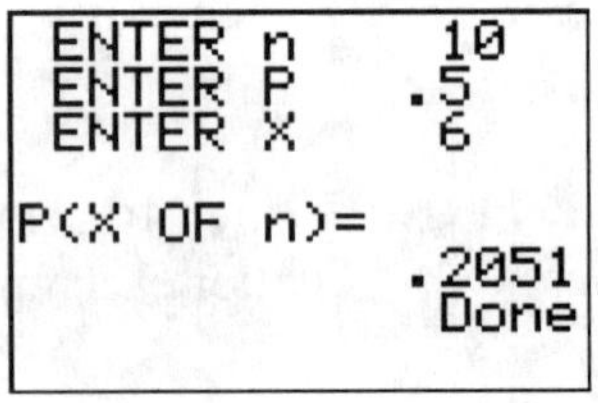

CLEAR or **ENTER**	**CLEAR** to use another function, or **ENTER** to run **BINOML83** again.

Solution: The probability of getting <u>exactly</u> 6 heads on 10 throws is .2051.
P(**x** = 6) = .2051

Example 2: Suppose that the coin is tossed 10 times. Find the probability that "heads" comes up at most 6 times.

[**BINOML82** is the corresponding program for the **TI-82**.]

NOTE: For this problem the probability of success "**p**" is .5, the number of trials "**n**" is 10, and the number of successes "**x**" is 0, 1, 2, 3, 4, 5, 6.

Steps

PRGM	Select the program menu.
BINOML83	Enter the number of the program corresponding to **BINOML83**.
ENTER	Start the program.
2	Select **SUM OF MANY**.

10	Enter the number of trials.
ENTER	(**n** = 10)
.5	Enter the probability of success on a
ENTER	single trial. (**p** = .5)
0	Enter the minimum number of successes.
ENTER	
6	Enter the maximum number of successes.
ENTER	

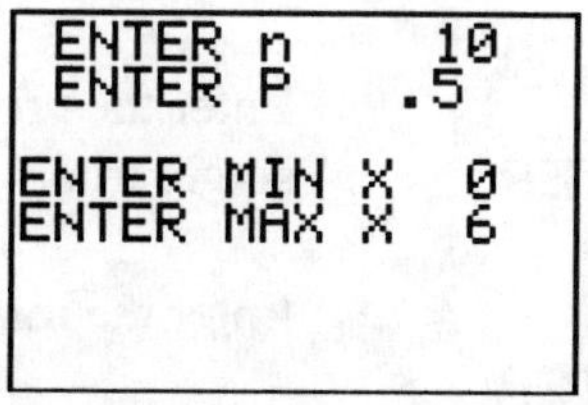

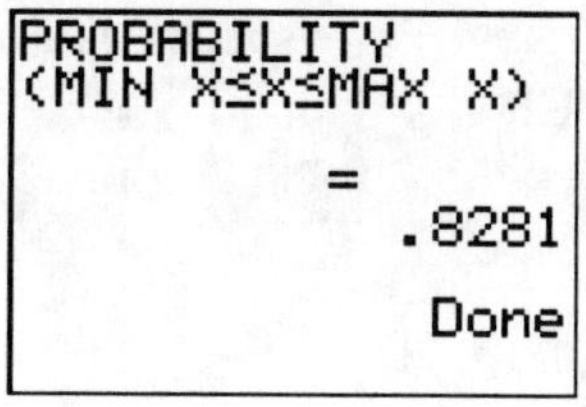

CLEAR or	**CLEAR** to use another function,
ENTER	or **ENTER** to run **BINOML83** again.

Solution: The probability of getting at most 6 heads on 10 throws is .8281.

$$P(0 \le \mathbf{x} \le 6) = .8281$$

Example 3: The probability that a given basketball player makes any single free throw is .78. Find the probability that if this player shoots 200 free throws, he will make at least 160, but less than 180 free throws.

[**BINOML82** is the corresponding program for the **TI-82**.]

NOTE: For this problem the probability of success "**p**" is .78, the number of trials "**n**" is 200, and the number of successes "**x**" is 160, 161, 162, … 179.

Steps

PRGM	Select the program menu.
BINOML83	Enter the number of the program corresponding to **BINOML83**.
ENTER	Start the program.
2	Select **SUM OF MANY**.

200	Enter the number of trials.
ENTER	(**n** = 200)
.78	Enter the probability of success on a
ENTER	single trial. (**p** = .78)
160	Enter the minimum number of successes.
ENTER	
179	Enter the maximum number of successes.
ENTER	

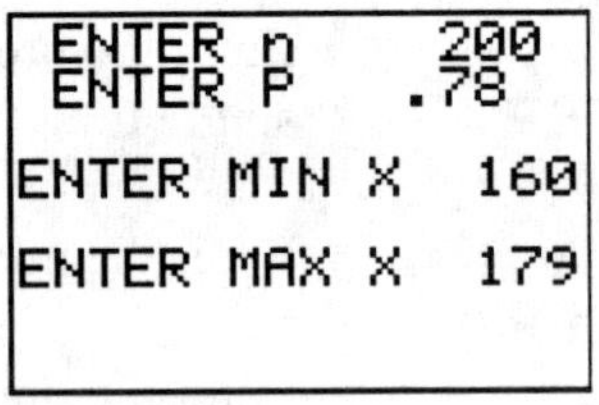

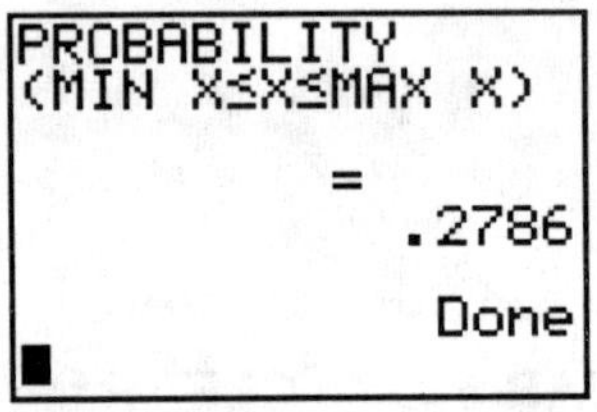

CLEAR or **ENTER**	**CLEAR** to use another function, or **ENTER** to run **BINOML83** again.

Solution: The probability that if this player shoots 200 free throws, he will make at least 160, but less that 180 free throws is .2786. $P(160 \le \mathbf{x} < 180) = .2786$

Mann: 5.6

Example 4: Suppose a coin is tossed 10 times. Let **x** be the number of heads that appear. Draw a ***probability distribution*** for this experiment.

[**BINOML82** is the corresponding program for the **TI-82**.]

Steps

PRGM	Select the program menu.
BINOML83	Enter the number of the program corresponding to **BINOML83**.
ENTER	Start the program.
3	Select **DISTRIBUTION**.

10	Enter the number of trials.
ENTER	(**n** = 10)
.5	Enter the probability of success on a
ENTER	single trial. (**p** = .5)

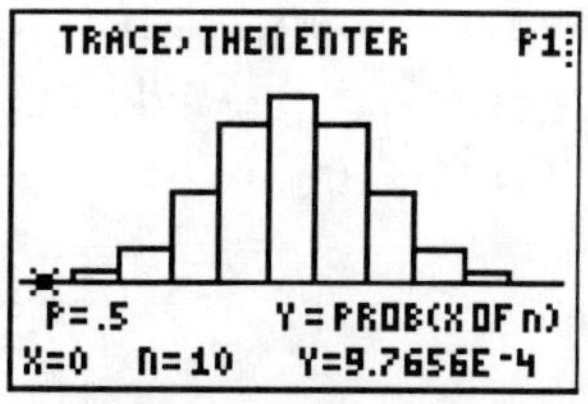

▶ ▶ ▶	Use the right arrow key and the screen will display the probabilities for specific values of **x**. For example, the probability that **x** = 3 out of 10 is .117.

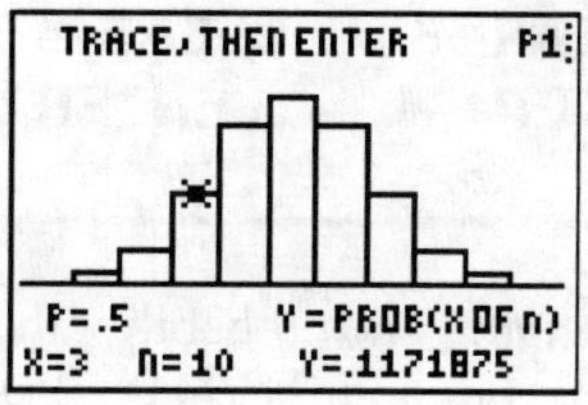

ENTER	End the program.
CLEAR or	**CLEAR** to use another function,
ENTER	or **ENTER** to run **BINOML83** again.

Poisson Probabilities Using the Program POISSN83

Example 1: A photocopy machine at a college breaks down an average of 8 times a month. Using the Poisson probability distribution, find the probability that during the next month the photocopy machine will break down exactly 9 times.

[**POISSN82** is the corresponding program for the **TI-82**.]

NOTE: In this problem the mean number of breakdowns per month "λ" is 8.

Steps

PRGM	Select the program menu.
POISSN83	Enter the number of the program corresponding to **POISSN83**.
ENTER	Start the program.
1	Select **SINGLE P(X)**.
8 **ENTER**	Enter the mean number of breakdowns per month. ($\lambda = 8$)
9 **ENTER**	Enter the specific number of breakdowns. ($\mathbf{x} = 9$)
CLEAR or **ENTER**	**CLEAR** to use another function, or **ENTER** to run **POISSN83** again.

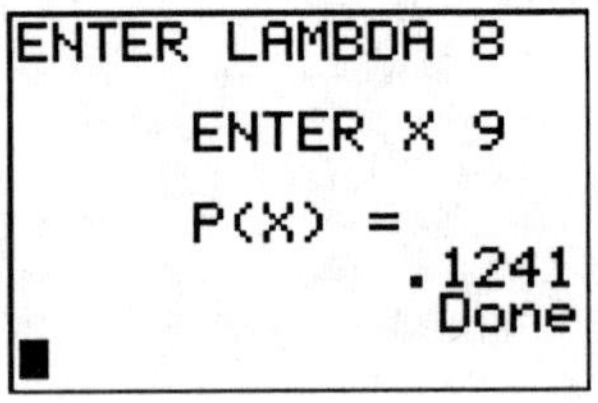

Solution: The probability that the photocopy machine will break down exactly 9 times in the next month is .1241. $P(\mathbf{x} = 9) = .1241$

Example 2: A photocopy machine at a college breaks down an average of 8 times a month. Using the Poisson probability distribution, find the probability that during the next month the photocopy machine will break down less than 8 times.

[**POISSN82** is the corresponding program for the **TI-82**.]

NOTE: In this problem the mean number of breakdowns per month "λ" is 8.

Steps

PRGM	Select the program menu.
POISSN83	Enter the number of the program corresponding to **POISSN83**.
ENTER	Start the program.
2	Select **SUM OF MANY**.
8 **ENTER**	Enter the mean number of breakdowns per month. ($\lambda = 8$)
0 **ENTER**	Enter the minimum value of **x**.
7 **ENTER**	Enter the maximum value of **x**.

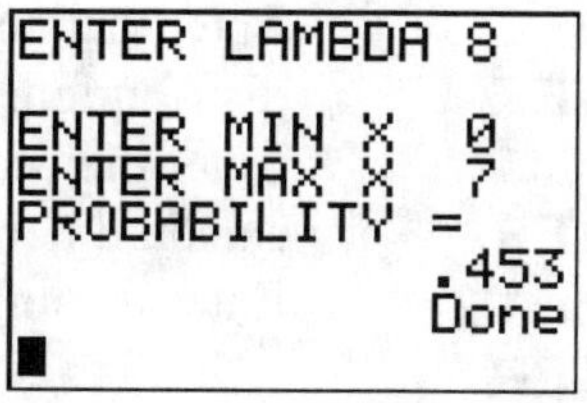

CLEAR or **ENTER**	**CLEAR** to use another function, or **ENTER** to run **POISSN83** again.

Solution: The probability that the photocopy machine will break down less than 8 times in the next month is .4530. $P(\mathbf{x} < 8) = P(0 \le \mathbf{x} \le 7) = .4530$

Probabilities in a Normal Distribution Using the Program **NORMAL83**

Example 1: If **z** has a standard normal distribution, find the probability that a value of **z** falls more than ½ of a standard deviation to the right of the mean.

[**NORMAL82** is the corresponding program for the **TI-82**.]

NOTE: In a standard normal distribution, $\mu = 0$ and $\sigma = 1$.

Steps

PRGM	Select the program menu.
NORMAL83	Enter the number of the program corresponding to **NORMAL83**.
ENTER	Start the program.
1	Select **P(X>A)**.
ENTER	
0	Enter the mean of the standard normal
ENTER	distribution. ($\mu = 0$)
1	Enter the standard deviation of the
ENTER	standard normal distribution. ($\sigma = 1$)
.5 (or 1/2)	Enter the minimum value of **z**.
ENTER	

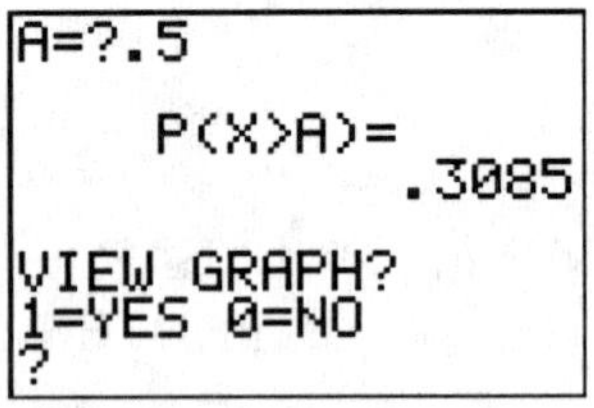

1	View the graph (after a brief wait).
ENTER	

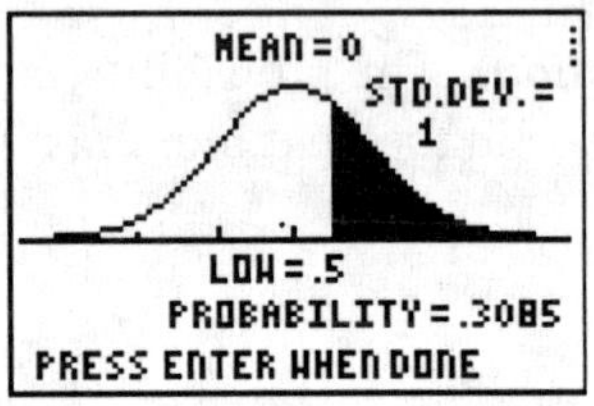

ENTER	End the program.
CLEAR or	**CLEAR** to use another function,
ENTER	or **ENTER** to run **NORMAL83** again.

Solution: The probability that **z** is greater than .5 is .3085. P(**z** > .5) = .3085

Example 2: A normal distribution has a mean of 10 and a standard deviation of 2. Find the probability that a randomly chosen value is less than 13.

[**NORMAL82** is the corresponding program for the **TI-82**.]

Steps

PRGM	Select the program menu.
NORMAL83	Enter the number of the program corresponding to **NORMAL83**.
ENTER	Start the program.
2	Select **P(X<B)**.
ENTER	
10	Enter the mean.
ENTER	
2	Enter the standard deviation.
ENTER	
13	Enter the maximum value of **x**.
ENTER	

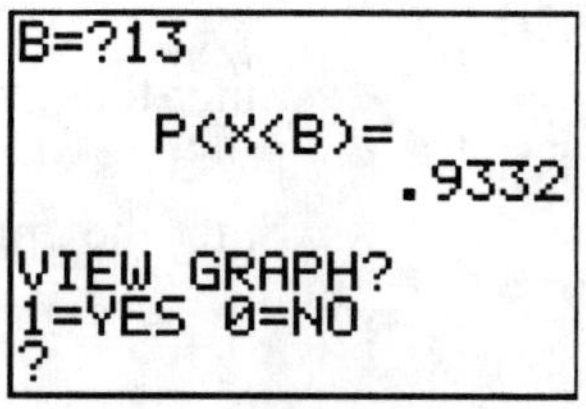

1	View the graph (after a brief wait).
ENTER	

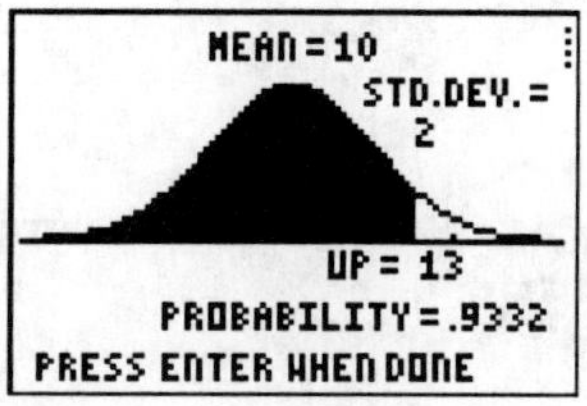

ENTER	End the program.
CLEAR or **ENTER**	**CLEAR** to use another function, or **ENTER** to run **NORMAL83** again.

Solution: The probability that **x** is less than 13 is .9332. P(**x** < 13) = .9332

Example 3: The normal distribution of "IQ" in the population has a mean of 100, and a standard deviation of 15. What is the probability that a randomly selected individual will have an IQ between 110 and 120?

[**NORMAL82** is the corresponding program for the **TI-82**.]

Steps

PRGM	Select the program menu.
NORMAL83	Enter the number of the program corresponding to **NORMAL83**.
ENTER	Start the program.
3	Select **P(A<X<B)**.
ENTER	
100	Enter the mean.
ENTER	
15	Enter the standard deviation.
ENTER	
110	Enter the lower limit, **A** = 110.
ENTER	
120	Enter the upper limit, **B** = 120.
ENTER	

```
A=?110
B=?120
  P(A<X<B)=
          .1613
VIEW GRAPH?
1=YES 0=NO
?
```

1	View the graph (after a brief wait).
ENTER	

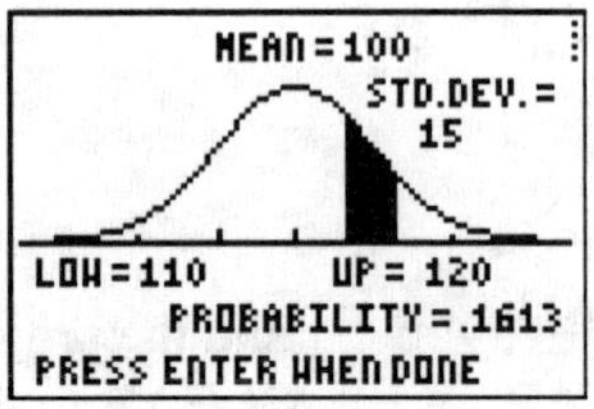

ENTER	End the program.
CLEAR or **ENTER**	**CLEAR** to use another function, or **ENTER** to run **NORMAL83** again.

Solution: The probability is .1613 that the IQ score of a randomly selected individual is between 110 and 120. $P(110 < \mathbf{x} < 120) = .1613$

Mann: 6.5

Finding z in a Normal Distribution Using the Program INVNOR83

Example 1: Find the **z** value for a standard normal distribution if the area in the left tail is .05.

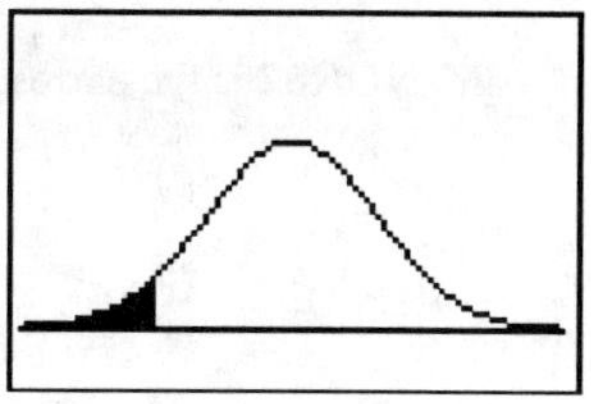

[**INVNOR82** is the corresponding program for the **TI-82**.]

Steps

PRGM	Select the program menu.
INVNOR83	Enter the number of the program corresponding to **INVNOR83**.
ENTER	Start program **INVNOR83**.
1	Select the area in left tail.

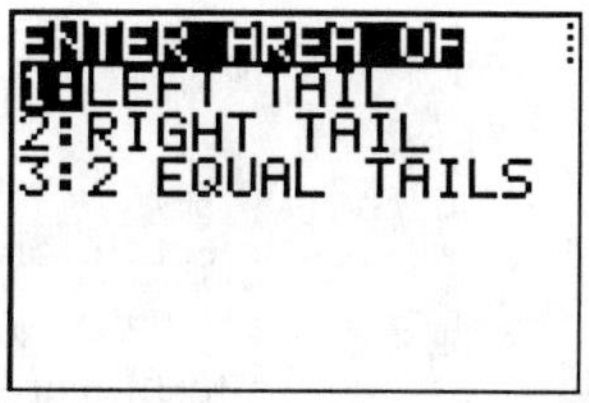

.05	Enter the area.
ENTER	

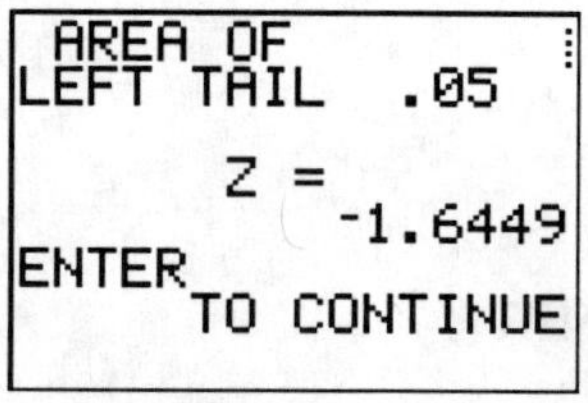

ENTER	
0	Select **0** as you only need a **z** score.
ENTER	
CLEAR or **ENTER**	**CLEAR** to use another function, or **ENTER** to run **INVNOR83** again.

Solution: The **z** value = -1.645, when the area in the left tail of the standard normal distribution is .05. P(**z** < -1.645) = .05

Example 2: Find the **z** value for a standard normal distribution if the area in the right tail is 0.125.

[**INVNOR82** is the corresponding program for the **TI-82**.]

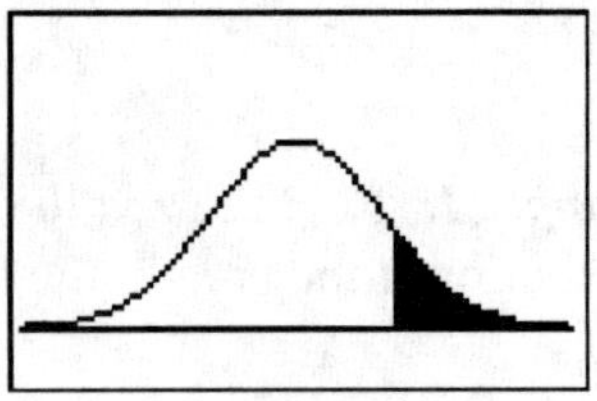

Steps

PRGM	Select the program menu.
INVNOR83	Enter the number of the program corresponding to **INVNOR83**.
ENTER	Start program **INVNOR83**.

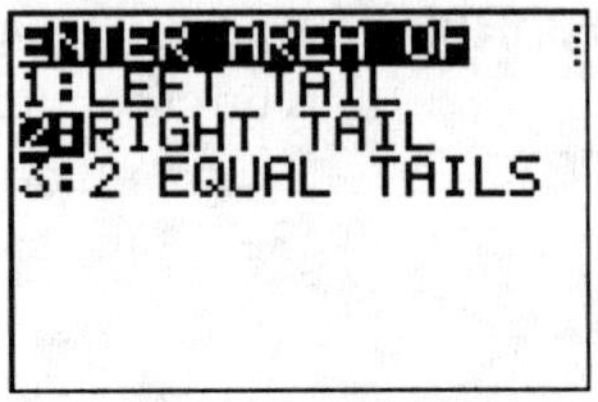

2	Select the area in <u>right</u> tail.
.125 **ENTER**	Enter the area.

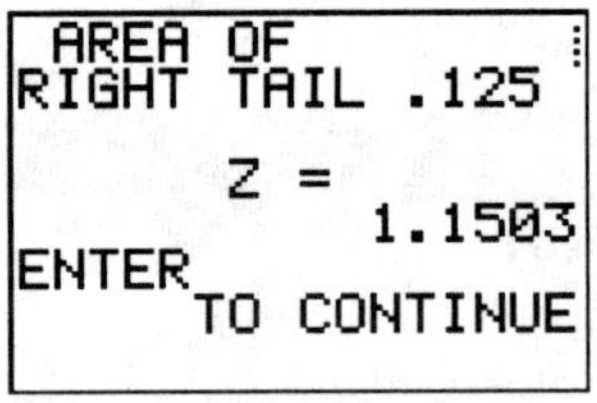

ENTER	
0	Select **0** as you need only a **z** score.
ENTER	
CLEAR or **ENTER**	**CLEAR** to use another function, or **ENTER** to run **INVNOR83** again.

Solution: The **z** value = 1.15, when the area in the <u>right</u> tail of the standard normal distribution is .125. P(**z** > 1.15) = .125

Example 3: Suppose that the height of adult males has a normal distribution with a mean of 70 inches and a standard deviation of 3 inches. How high should a door be so that 98% of adult males can pass through without ducking?

[**INVNOR82** is the corresponding program for the **TI-82**.]

Steps

PRGM	Select the program menu.
INVNOR83	Enter the number of the program corresponding to **INVNOR83**.
ENTER	Start program **INVNOR83**.
1	Select the area in left tail.
.98	Enter the area.
ENTER	

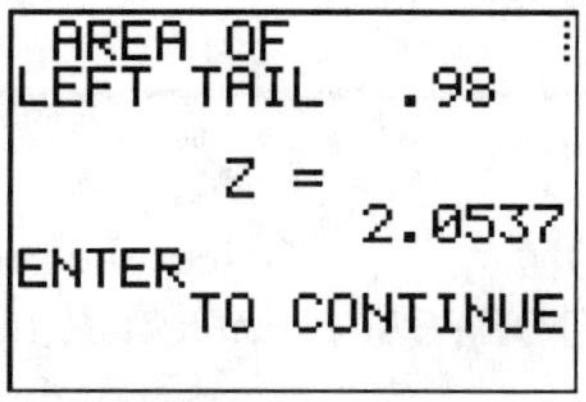

ENTER	
1	Select **1** as you need a value of **x**.
ENTER	
70	Enter the mean.
ENTER	
3	Enter the standard deviation.
ENTER	

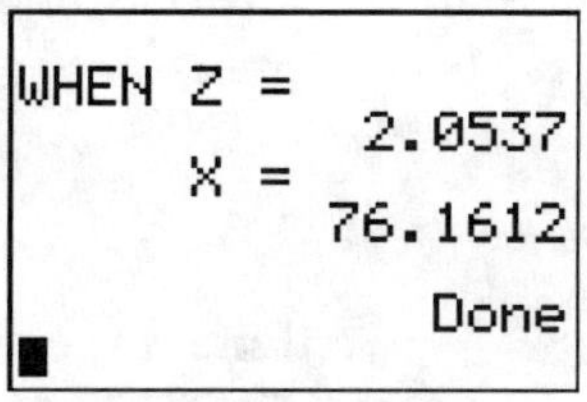

CLEAR or **ENTER**	**CLEAR** to use another function, or **ENTER** to run **INVNOR83** again.

Solution: The height of a door should be 76.16 inches to permit 98% of adult men to pass through without ducking. P(**x** < 76.16) = .98

Goodness-of-Fit Test Using the Program GOODFT83

Example: Suppose that to grade on a curve means to give 10% A's, 20% B's, 40%C's, 20% D's and 10% F's. A teacher actually gave 10 A's, 18 B's, 35 C's, 12 D's and 5 F's. At the 5 % significance level would you say that these grades are a "good fit" to a curve.

[**GOODFT82** is the corresponding program for the **TI-82**.]

FIRST: Set up problem: H_0: The grades are a good fit to a curve
H_1: The grades are not a good fit to a curve

A goodness-of-fit test uses the χ^2 distribution and is always right-tailed. This problem has 4 degrees of freedom. The critical value for this problem is 9.488.

Steps

PRGM	Select the program menu.
GOODFT83	Enter the number of the program corresponding to **GOODFT83**.
ENTER	Start program **GOODFT83**.
5	Enter the number of outcomes.
ENTER	
Enter Observed Values	Enter the observed values. Press **ENTER** after each entry.

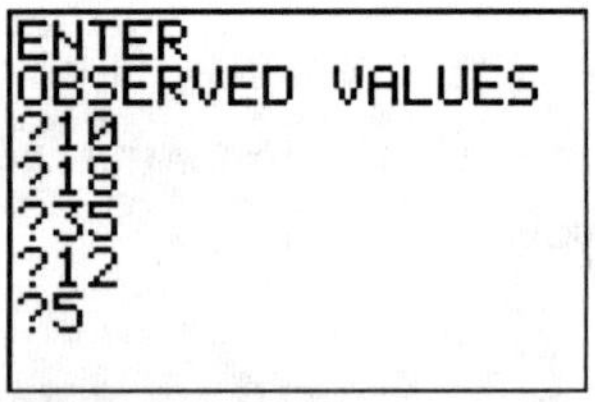

Enter Expected Values	Enter the expected values. Press **ENTER** after each entry.

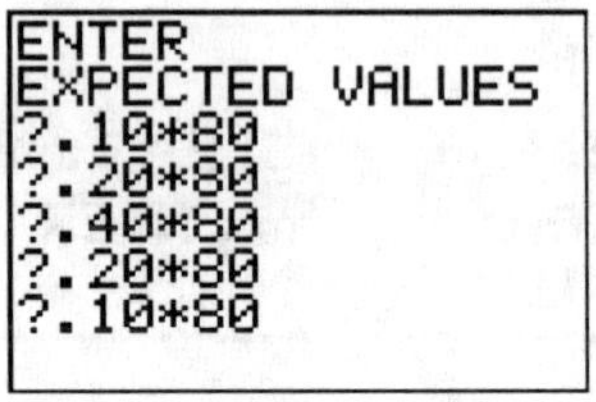

[**TI-82** screens differ.]

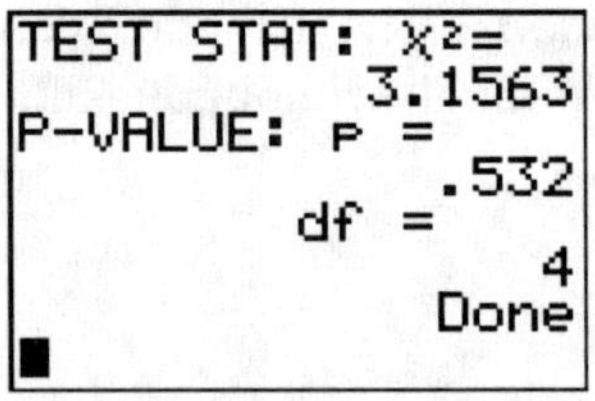

Decision: The test statistic is 3.1563. The critical value is 9.488 for this right-tailed test. The significance level is 5% while the p-value is .532. The decision is to not reject H_0. The grades the teacher gave are a good fit to a curve.

TI-82 Program Notes

The **TI-82** cannot perform all of the functions of the **TI-83**. Specifically, the **TI-82** does not have a Distribution (**DISTR**) key or a **TESTS** menu when you press the **STAT** key. The programs described in this section emulate many **TI-83** functions, but not all of them. Short descriptions of these **TI-82** programs are given below.

ANOVA : Computes the test statistic and the degrees of freedom for a one-way ANOVA test. The data from the samples should be entered into the lists before starting the program. The data from the first sample should be in List 1, the data from the second sample in List 2, and so on. The program will give you a choice of 2 to 5 samples.

BINOML82 : See page 86 for a detailed description. This program is also available in a **TI-83** version.

CHISQ : Computes the test statistic, the degrees of freedom and the p-value for a test of independence or a test for homogeneity. In order to compute the p-value, the program calls upon the following other programs as subroutines: **ZPVALUE**, **ZCHIAREA**, and **ZGAMMA**. These subroutines must be in the calculator to compute the p-value.

GOODFT82 : See page 98 for a detailed description. This program is also available in a **TI-83** version. In order to compute the p-value, the program calls upon the following other programs as subroutines: **ZPVALUE**, **ZCHIAREA**, and **ZGAMMA**. These subroutines must be in the calculator to compute the p-value.

HYPTESTB : Computes the test statistic and the p-value for the hypothesis test about the slope of the regression line. The independent variable data must be in List 1 and the dependent variable data must be in List 2 before starting the program. In order to compute the p-value, the program calls up the following subroutines: **ZPVALUE**, **ZTAREA**, and **ZGAMMA**. These subroutines must be in the calculator to compute the p-value.

INVNOR82 : See page 95 for a detailed description. This program is also available in a **TI-83** version.

NORMAL82 : See page 92 for a detailed description. This program is also available in a **TI-83** version.

ONEPOPCI : Computes the confidence interval based on a sample from one population. The program can do confidence intervals for a mean or a proportion. Either raw data or summary statistics can be used for the mean. If using raw data, the data must be entered into List 1 before starting the program.

ONEPOPHT : Computes the test statistic and the p-value for a sample from one population. The program can find these values for both means and proportions. Either summary statistics or raw data can be used for the mean. If using raw data, the data from the sample must be entered in List 1 before starting the program. In order to compute the p-value, the program calls up the following subroutines: **ZPVALUE**, **ZTAREA**, and **ZGAMMA**. These subroutines must be in the calculator to compute the p-value.

POISSN82 : See page 90 for a detailed description. This program is also available in a **TI-83** version.

PROBDIST : Computes the mean and standard deviation of a discrete probability distribution. Since this program is likely to be the first program used on the **TI-82**, detailed instructions are given on page 85.

RANDOMNO : Generates random positive integers. Enter the largest positive integer to be generated. Press the [ON] key to stop the program.

TWOPOP : Calculates test statistics and confidence intervals for the difference between the population means or the difference between the population proportions based on samples from two populations. The first screen gives a choice of computing the confidence interval, the test statistic or both. The second screen gives a choice of means or proportions. If "**MEANS**" is the choice, the next screen asks for the source of the data. If using raw data, the data sample from Population 1 must be entered into List 1 and the data sample from Population 2 must be entered into List 2 before starting the program. In order to compute the p-value, the program calls up the following subroutines: **ZPVALUE**, **ZTAREA**, and **ZGAMMA**. These subroutines must be in the calculator to compute the p-value.

ZPVALUE : This program is a subroutine used by the programs **CHISQ**, **GOODFT82**, **HYPTESTB**, **ONEPOPHT**, and **TWOPOP**. **ZPVALUE** calls up the subroutines **ZTAREA**, **ZCHIAREA**, and **ZGAMMA**. The subroutines must be in the calculator for the **ZPVALUE** program to run.

Appendix

Program **BINOML83.83P**

```
ClrHome:ClrList L5,L6
Menu("BINOMIAL PRQB.","SINGLE P(X)",1,"SUM OF MANY",2,"DISTRIBUTION",3)
Lbl 1
Input (" ENTER n      ",N)
If fPart(N)≠Ø or N<Ø
Goto 7
Input (" ENTER P     ",P)
If P<Ø or P>1
Goto 7
Input (" ENTER X      ",X)
If fPart(X)≠Ø or X<Ø or X>N
Goto 7
binompdf(N,P,X)→K
Disp ""
Disp "P(X OF n)=",round(K,4)
Stop
Lbl 2
Input (" ENTER n      ",N)
If fPart(N)≠Ø or N<Ø
Goto 7
Input (" ENTER P     ",P)
If P<Ø or P>1
Goto 7
Disp ""
Input ("ENTER MIN X  ",A)
If A<Ø or A>N
Goto 7
If fPart(A)≠Ø
Goto 7
Input ("ENTER MAX X  ",B)
If fPart(B)≠Ø
Goto 7
If B<Ø or B>N
Goto 7
If (B-A)<Ø
Goto 7
binomcdf(N,P,B)-binomcdf(N,P,(A-1))→S
ClrHome
Disp "PROBABILITY"
Disp "(MIN X≤X≤MAX X)"
Disp ""
Disp "           =",round(S,4)
Disp ""
Stop
Lbl 3
FnOff :PlotsOff
Disp ""
Input (" ENTER n      ",N)
If fPart(N)≠Ø or N<Ø or N=Ø
Goto 7
If N>46
Goto 8
Disp ""
Input (" ENTER P     ",P)
If P<Ø or P>1
Goto 7
StoreGDB 9
ExprOff
For(X,Ø,N)
binompdf(N,P,X)→L6(X+1)
End
For(H,Ø,N):H→L5(H+1)
End
Plot1(Scatter,L5,L6,·)
Plot2(Histogram,L5,L6)
-.5→Xmin
N+.5→Xmax
1→Xscl
-.5*max(L6)→Ymin
1.4*max(L6)→Ymax
AxesOff
DispGraph
Text(57,21,"N=")
Text(57,3Ø,N)
Text(49,5,"P=")
Text(49,14,round(P,3))
Text(49,45,"Y = PROB(X OF n)")
Text(Ø,7,"TRACE, THEN ENTER")
Trace
RecallGDB 9
ClrList L5,L6
PlotsOff :FnOn
AxesOn:ExprOn
Float:ClrHome:Stop
Lbl 7
Disp " "
Disp " INVALID ENTRY"
Disp " "
Disp "PLEASE TRY AGAIN"
Stop
Lbl 8
ClrHome
Disp "n MUST BE ≤ 46"
Disp "FOR THIS PROGRAM"
Disp "TO WORK, SORRY!"
Disp " "
Disp "PLEASE TRY AGAIN"
Stop
```

Program **NORMAL83.83P**

```
StoreGDB 9
PlotsOff :Func:Connected:ClrHome
FnOff
Disp "NORMAL"
Disp "    PROBABILITY"
Disp "1:P(X>A)"
Disp "2:P(X<B)"
Disp "3:P(A<X<B)"
Disp ""
Input ("ENTER CHOICE  ",C)
Disp "":
Input ("ENTER MEAN  ",M)
Disp " "
Disp "ENTER "
Disp " STANDARD"
Input ("DEVIATION   ",S)
If S=Ø or S<Ø
Goto 9
If C=1
Then
Disp "":Disp "P(X>A)"
Prompt A
A→L
M+6S→U
Disp ""
Disp "    P(X>A)="
End
If C=2
Then
Disp "":Disp "P(X<B)"
Prompt B
B→U
M-6S→L
Disp ""
Disp "    P(X<B)="
End
If C=3
Then
Disp ""
Disp "P(A<X<B)"
Prompt A:Prompt B:
A→L:B→U
If (U-L)≤Ø
Goto 9
Disp ""
Disp "  P(A<X<B)="
End
normalcdf(L,U,M,S)→P
If P<Ø
Ø→P
Disp round(P,4)
If C≠3
Disp ""
Disp "VIEW GRAPH?"
Disp "1=YES Ø=NO"
Input E
If E=Ø:Goto 8
⁻3.5→Xmin
3.5→Xmax
⁻.36→Ymin
.56→Ymax
AxesOff
Line(⁻3.5,Ø,3.5,Ø)
Line(⁻1,Ø,⁻1,.Ø2)
Line(⁻2,Ø,⁻2,.Ø1)
Line(1,Ø,1,.Ø2)
Line(2,Ø,2,.Ø1)
Line(Ø,Ø,Ø,.Ø2)
ShadeNorm((L-M)/S,(U-M)/S,Ø,1)
Text(48,26,"PROBABILITY =")
Text(48,76,round(P,4))
Text(1,34,"MEAN =")
Text(1,56,round(M,4))
Text(9,6Ø,"STD.DEV. =")
Text(16,76,round(S,4))
If C=1
Goto 6
If C=2
Goto 5
Text(4Ø,Ø3,"LOW =")
Text(4Ø,21,round(L,4))
Lbl 5
Text(4Ø,54,"UP =")
Text(4Ø,68,round(U,4))
Text(57,3,"PRESS  ENTER  WHEN DONE")
Goto 7
Lbl 6
Text(4Ø,33,"LOW =")
Text(4Ø,51,round(L,4))
Text(57,3,"PRESS  ENTER  WHEN DONE")
Lbl 7
Pause
Lbl 8
RecallGDB 9:ClrHome
FnOn :Float
Stop
Lbl 9
Disp ""
Disp " INVALID ENTRY"
Disp ""
Disp "PLEASE TRY AGAIN"
RecallGDB 9
FnOn :Float
Stop
```

Program **INVNOR83.83P**

```
Menu("ENTER AREA OF","LEFT TAIL",1,"RIGHT TAIL",2,"2 EQUAL TAILS",3)
Lbl 1
ClrHome
Disp " "
Disp "ENTER "
Disp " AREA OF "
Input ("LEFT TAIL  ",A)
If A≤Ø or A≥1
Goto 11
invNorm(A)→Z
Disp ""
Disp "       Z =",round(Z,3)
Disp "ENTER "
Disp "     TO CONTINUE"
1→W
Goto 2Ø
Lbl 2
ClrHome
Disp ""
Disp "ENTER "
Disp " AREA OF "
Input ("RIGHT TAIL ",A)
If A≤Ø or A≥1
Goto 11
-1*(invNorm(A))→Z
Disp ""
Disp "       Z =",round(Z,3)
Disp "ENTER "
Disp "     TO CONTINUE"
2→W
Goto 2Ø
Lbl 3
ClrHome
Disp "ENTER TOTAL"
Disp " AREA OF"
Input ("TWO TAILS  ",A)
If A≤Ø or A>1:Goto 11
invNorm(A/2)→L:-L→U
Disp ""
Disp " UPPER Z =",round(U,3)
Disp " LOWER Z =",round(L,3)
U→Z
3→W
Disp ""
Disp "ENTER "
Disp "     TO CONTINUE"
Goto 2Ø
Lbl 11
Disp ""
Disp " INVALID ENTRY "
Disp ""
Disp "PLEASE TRY AGAIN"
Stop
Lbl 2Ø
Pause
ClrHome
Disp "ANSWERS IN TERMS"
Disp "  OF  X ?"
Disp ""
Input ("1=YES  Ø=NO   ",E)
If E=1
Goto 21
If W=3
Goto 33
ClrHome
Disp " "
Disp "       Z =",round(Z,3)
Disp " "
Stop
Lbl 21
Disp " "
Input ("ENTER MEAN ",M)
Disp " "
Disp " "
Disp " "
Disp "ENTER "
Disp " STANDARD"
Input ("DEVIATION  ",S)
If S=Ø or S<Ø
Goto 11
ClrHome
Disp " ":Disp " "
If W=3
Goto 32
S*Z+M→X
Disp "WHEN Z =",round(Z,3)
Disp "     X =",round(X,3)
Disp " "
Stop
Lbl 32
S*L+M→X
Disp " LOWER Z =",round(L,3)
Disp " "
Disp " LOWER X =",round(X,3)
Disp "ENTER "
Disp "     TO CONTINUE"
Pause
U→Z
S*U+M→X
Disp ""
Disp " UPPER Z =",round(U,3)
Disp " "
Disp " UPPER X =",round(X,3)
Disp " "
Stop
Lbl 33
ClrHome
Stop
```

Program **BINOML82.82P**

```
:ClrHome:ClrList L5,L6
:Menu("BINOMIAL PRQB.","SINGLE P(X)",1,"SUM OF MANY",2,"DISTRIBUTION",3)
:Lbl 1
:Input (" ENTER n     ",N)
:If fPart N≠Ø or N<Ø
:Goto 7
:Input (" ENTER P    ",P)
:If P<Ø or P>1
:Goto 7
:Input (" ENTER X     ",X)
:If fPart X≠Ø or X<Ø or X>N
:Goto 7
:N nCr X*P^X*(1-P)^(N-X)→K
:Disp " ":Disp "P(X OF
n)=",round(K,4)
:Stop
:Lbl 2
:Input (" ENTER n     ",N)
:If fPart N≠Ø or N<Ø
:Goto 7
:Input (" ENTER P    ",P)
:If P<Ø or P>1
:Goto 7
:Disp ""
:Input ("ENTER MIN X  ",A)
:If A<Ø or A>N
:Goto 7
:If fPart A≠Ø
:Goto 7
:Input ("ENTER MAX X  ",B)
:If fPart B≠Ø
:Goto 7
:If B<Ø or B>N
:Goto 7
:If (B-A)<Ø
:Goto 7
:Ø→S
:For(X,A,B)
:N nCr X*P^X*(1-P)^(N-X)→T
:T+S→S
:End
:Disp "PROBABILITY"
:Disp "MIN X≤X≤MAX X"
:Disp ""
:Disp "           =",round(S,4)
:Disp ""
:Stop
:Lbl 3
:FnOff :PlotsOff
:Input (" ENTER n     ",N)
:If fPart N≠Ø or N<Ø or N=Ø
:Goto 7
:If N>46
:Goto 8
:Input (" ENTER P    ",P)
:If P<Ø or P>1
:Goto 7
:StoreGDB GDB6
:For(X,Ø,N)
:N nCr X*P^X*(1-P)^(N-X)→L6(X+1)
:End
:For(H,Ø,N):H→L5(H+1)
:End
:round(L6*1ØØ,Ø)→L6
:Plot1(Scatter,L5,L6,·)
:Plot2(Histogram,L5,L6)
:-.5→Xmin
:N+.5→Xmax
:1→Xscl
:-.5*max(L6)→Ymin
:1.4*max(L6)→Ymax
:AxesOff
:DispGraph
:Text(57,21,"n=")
:Text(57,3Ø,N)
:Text(49,5,"P=")
:Text(49,14,round(P,2))
:Text(49,44,"Y = PROB(X OF n)")
:Text(57,67,"/1ØØ")
:Text(Ø,7,"TRACE, THEN ENTER")
:Trace
:RecallGDB GDB6
:ClrList L5,L6
:PlotsOff :FnOn
:AxesOn
:Float:ClrHome:Stop
:Lbl 7
:Disp ""
:Disp " INVALID ENTRY"
:Disp ""
:Disp "PLEASE TRY AGAIN"
:Stop
:Lbl 8
:ClrHome
:Disp "n MUST BE ≤ 46"
:Disp "FOR THIS PROGRAM"
:Disp "TO WORK, SORRY!"
:Disp ""
:Disp "PLEASE TRY AGAIN"
:Stop
```

Program **NORMAL82.82P**

```
:StoreGDB GDB6
:FnOff
:PlotsOff :Func:Connected:ClrHome
:"(√(2π))⁻¹*e^(⁻.5*(X)² )"→Y₀
:FnOff Ø
:Disp "NORMAL"
:Disp "     PROBABILITY"
:Disp "1:P(X>A)"
:Disp "2:P(X<B)"
:Disp "3:P(A<X<B)"
:Disp ""
:Input ("ENTER CHOICE  ",C)
:Disp ""
:Input ("ENTER MEAN  ",M)
:Disp ""
:Disp "ENTER"
:Disp " STANDARD"
:Input ("DEVIATION   ",S)
:If S=Ø or S<Ø
:Goto 9
:If C=1
:Then
:Disp ""
:Disp "P(X>A)"
:Prompt A
:A→L
:M+6S→U
:Disp ""
:Disp "    P(X>A)="
:End
:If C=2
:Then
:Disp ""
:Disp "P(X<B)"
:Prompt B
:B→U
:M-6S→L
:Disp ""
:Disp "    P(X<B)="
:End
:If C=3
:Then
:Disp ""
:Disp "P(A<X<B)"
:Prompt A:Prompt B:
:A→L:B→U
:If (U-L)≤Ø
:Goto 9
:Disp ""
:Disp "  P(A<X<B)="
:End
:(L-M)/S→R
:(U-M)/S→T
:fnInt(Y₀,X,R,T,.ØØØ1)→P
:If P<Ø
:Ø→P
:Disp round(P,4)
:If C≠3
:Disp ""
:Disp "VIEW GRAPH?"
:Disp "1=YES Ø=NO"
:Input E
:If E=Ø:Goto 8
:⁻3.5→Xmin
:3.5→Xmax
:⁻.36→Ymin
:.56→Ymax
:AxesOff
:Line(⁻3.5,Ø,3.5,Ø)
:Line(⁻1,Ø,⁻1,.Ø2)
:Line(⁻2,Ø,⁻2,.Ø1)
:Line(1,Ø,1,.Ø2)
:Line(2,Ø,2,.Ø1)
:Line(Ø,Ø,Ø,.Ø2)
:Shade(Ø,Y₀,1,R,T)
:Text(48,29,"PROBABILITY =")
:Text(48,76,round(P,4))
:Text(1,34,"MEAN =")
:Text(1,59,round(M,4))
:Text(9,6Ø,"STD.DEV. =")
:Text(16,76,round(S,4))
:If C=1
:Goto 6
:If C=2
:Goto 5
:Text(4Ø,Ø3,"LOW =")
:Text(4Ø,23,round(L,4))
:Lbl 5
:Text(4Ø,54,"UP =")
:Text(4Ø,69,round(U,4))
:Text(57,3,"PRESS  ENTER  WHEN
DONE")
:Goto 7
:Lbl 6
:Text(4Ø,33,"LOW =")
:Text(4Ø,53,round(L,4))
:Text(57,3,"PRESS  ENTER  WHEN
DONE")
:Lbl 7
:Pause
:Lbl 8
:RecallGDB GDB6
:ClrHome:FnOn
:Float:Stop
:Lbl 9
:Disp ""
:Disp " INVALID ENTRY"
:Disp ""
:Disp "PLEASE TRY AGAIN"
:RecallGDB GDB6
:FnOn
:Float:Stop
```

Program **INVNOR82.82P**

```
:2.515517→A:.802853→B
:.010328→C:1.432788→D
:.189269→E:.001308→F
:Menu("ENTER AREA OF","LEFT TAIL",1,"RIGHT TAIL",2,"2 EQUAL TAILS",3)
:Lbl 1
:⁻1→H
:ClrHome
:Disp " "
:Disp "ENTER "
:Disp " AREA OF"
:Input ("LEFT TAIL  ",G)
:1→W
:Goto 4
:Lbl 2
:1→H
:ClrHome
:Disp " "
:Disp "ENTER "
:Disp " AREA OF"
:Input ("RIGHT TAIL ",G)
:2→W
:Lbl 4
:If G≤0 or G≥1:Goto 9
:If G>.5 and G<1:⁻H→H
:If G>.5 and G<1:1-G→G
:√ln (1/G² )→T
:(T-(A+B*T+C*T^2)/(1+D*T+E*T^2+F*T^3))*H→Z
:Disp " "
:Disp "       Z =",round(Z,2)
:Disp "ENTER"
:Disp "     TO CONTINUE"
:Goto 5
:Lbl 3
:3→W:ClrHome
:Disp " "
:Disp "ENTER TOTAL"
:Disp " AREA OF"
:Input ("TWO TAILS  ",G)
:If G≤0 or G>1:Goto 9
:.5*G→G
:√ln (1/G² )→T
:(T-(A+B*T+C*T^2)/(1+D*T+E*T^2+F*T^3))→Z
:⁻Z→L:Z→U
:Disp ""
:Disp " UPPER Z =",round(U,2)
:Disp " LOWER Z =",round(L,2)
:Disp ""
:Disp "ENTER "
:Disp "     TO CONTINUE"
:Goto 5
:Lbl 9
:Disp ""
:Disp " INVALID ENTRY "
:Disp ""
:Disp "PLEASE TRY AGAIN"
:Stop
:Lbl 5
:Pause
:ClrHome
:Disp "ANSWERS IN TERMS"
:Disp "  OF  X ?"
:Disp ""
:Input ("1=YES  0=NO   ",J)
:If J=1
:Goto 6
:If W=3
:Goto 8
:ClrHome
:Disp " "
:Disp "       Z =",round(Z,2)
:Disp " "
:Stop
:Lbl 6
:Disp " "
:Input ("ENTER MEAN ",M)
:Disp " "
:Disp " "
:Disp " "
:Disp "ENTER "
:Disp " STANDARD"
:Input ("DEVIATION  ",S)
:If S=0 or S<0
:Goto 9
:ClrHome
:Disp " ":Disp " "
:If W=3
:Goto 7
:S*Z+M→X
:Disp "WHEN Z =",round(Z,2)
:Disp "       X =",round(X,3)
:Disp " "
:Stop
:Lbl 7
:S*L+M→X
:Disp " LOWER Z =",round(L,2)
:Disp " "
:Disp " LOWER X =",round(X,3)
:Disp "ENTER "
:Disp "     TO CONTINUE"
:Pause
:U→Z
:S*U+M→X
:Disp ""
:Disp " UPPER Z =",round(U,2)
:Disp " "
:Disp " UPPER X =",round(X,3)
:Disp " "
:Stop
:Lbl 8
:ClrHome:Stop
```

Program **POISSN83.83P**

```
ClrHome
Menu("POISSON PROB.","SINGLE
P(X)",1,"SUM OF MANY",2)
Lbl 1
Input "ENTER LAMBDA ",L
Disp ""
Input "      ENTER X ",X
(L^X*e^(-L))/(X!)→A
Disp ""
Disp "      P(X) = ",round(A,4)
Stop
Lbl 2
Ø→C
Input "ENTER LAMBDA ",L
Disp ""
Input "ENTER MIN X  ",A
Input "ENTER MAX X  ",B
For(X,A,B)
(L^X*e^(-L))/(X!)→D
D+C→C
End
Disp "PROBABILITY =",round(C,4)
Stop
```

Program **GOODFT83.83P**

```
ClrHome
ClrList L4,L5,L6
Disp "GOODNESS-OF-FIT"
Disp "TEST"
Disp " "
Disp "ENTER NUMBER"
Input " OF OUTCOMES ",N
ClrHome
Disp "ENTER "
Disp "OBSERVED VALUES"
For(I,1,N)
Input X
X→L4(I)
End
ClrHome
Disp "ENTER "
Disp "EXPECTED VALUES"
For(I,1,N)
Input X
X→L5(I)
End
For(I,1,N)
(L4(I)-L5(I))²/L5(I)→L6(I)
End
ClrHome
Disp "TEST STAT: χ² = "
sum(L6)→S
Disp round(S,4)
Disp "P-VALUE: p = "
1-χ²cdf(Ø,S,N-1)→P
Disp round(P,4)
Disp "        df ="
Disp (N-1)
Stop
```

Program **ANOVA.82P**

```
:ClrHome
:Disp "ANOVA"
:Disp "ENTER DATA"
:Disp "IN LISTS BEFORE"
:Disp "STARTING PROGRAM"
:Disp " "
:Disp "CONTINUE?"
:Input "1=YES,0=NO",S
:If S=0:Stop
:Menu("NO. OF SAMPLES?","CHOOSE
ONE",1,"TWO",2,"THREE",3,"FOUR",4,"
FIVE",5)
:Lbl 1:Stop
:Lbl 2
:sum L1→A
:dim L1→B
:sum L2→C
:dim L2→D
:sum (L1²)+sum (L2²)→G
:2→K
:B+D→N
:ClrHome
:Disp "D OF F NUM =",K-1
:Disp "D OF F DEN =",N-K
:Disp "PRESS ENTER"
:Pause
:((A²/B+C²/D)-(A+C)²/N)→H
:Disp "MSB =",H
:((G-A²/B-C²/D)/(N-K)→I
:Disp "MSW =",I
:H/I→T
:Disp "TEST STAT =",T
:Stop
:Lbl 3
:sum L1→A
:dim L1→B
:sum L2→C
:dim L2→D
:sum L3→E
:dim L3→F
:sum (L1²)+sum (L2²)+sum (L3²)→G
:3→K
:B+D+F→N
:ClrHome
:Disp "D OF F NUM =",K-1
:Disp "D OF F DEN =",N-K
:Disp "PRESS ENTER"
:Pause
:((A²/B+C²/D+E²/F)-(A+C+E)²/N)/2→H
:Disp "MSB =",H
:((G-A²/B-C²/D-E²/F)/(N-K)→I
:Disp "MSW =",I
:H/I→T
:Disp "TEST STAT =",T
:Stop
:Lbl 4
:sum L1→A
:dim L1→B
:sum L2→C
:dim L2→D
:sum L3→E
:dim L3→F
:sum (L1²)+sum (L2²)+sum (L3²)+sum
(L4²)→G
:4→K
:B+D+F+dim L4→N
:ClrHome
:Disp "D OF F NUM =",K-1
:Disp "D OF F DEN =",N-K
:Disp "PRESS ENTER"
:Pause
:((A²/B+C²/D+E²/F+(sum L4)²/dim
L4)-(A+C+E+sum L4)²/N)/3→H
:Disp "MSB =",H
:((G-A²/B-C²/D-E²/F-(sum L4)²/dim
L4)/(N-K)→I
:Disp "MSW =",I
:H/I→T
:Disp "TEST STAT =",T
:Stop
:Lbl 5
::sum L1→A
:dim L1→B
:sum L2→C
:dim L2→D
:sum L3→E
:dim L3→F
:sum (L1²)+sum (L2²)+sum (L3²)+sum
(L4²)+sum (L5²)→G
:5→K
:B+D+F+dim L4+dim L5→N
:ClrHome
:Disp "D OF F NUM =",K-1
:Disp "D OF F DEN =",N-K
:Disp "PRESS ENTER"
:Pause
:((A²/B+C²/D+E²/F+(sum L4)²/dim
L4+(sum L5)²/dim L5)-(A+C+E+sum
L4+sum L5)²/N)/4→H
:Disp "MSB =",H
:((G-A²/B-C²/D-E²/F-(sum L4)²/dim
L4-(sum L5)²/dim L5)/(N-K)→I
:Disp "MSW =",I
:H/I→T
:Disp "TEST STAT =",T
:Stop
```

Program **CHISQ.82P**

```
:ClrHome
:Disp "FINDS THE TEST"
:Disp "STATISTIC FOR"
:Disp "INDEPENDENCE OR"
:Disp "HOMOGENEITY"
:Disp " "
:Input "NO ROWS = ",R
:Input "NO COLMS = ",C
:{R+1,C+1}→dim [A]
:{R+1,C+1}→dim [E]
:{R,1}→dim [B]
:{1,C}→dim [C]
:{R,C}→dim [D]
:For(I,1,R)
:For(J,1,C)
:0→[A](I,J)
:End
:End
:For(I,1,R)
:ClrHome
:Disp "ROW",I
:For(J,1,C)
:Disp "COLUMN ",J
:Prompt X
:X→[A](I,J)
:End
:End
:For(I,1,R)
:[A](I,1)→[B](I,1)
:End
:For(I,1,R)
:For(J,2,C)
:[B](I,1)+[A](I,J)→[B](I,1)
:End
:End
:For(J,1,C)
:[A](1,J)→[C](1,J)
:End
:For(J,1,C)
:For(I,2,R)
:[C](1,J)+[A](I,J)→[C](1,J)
:End
:End
:0→T
:For(I,1,R)
:T+[B](I,1)→T
:End
:For(I,1,R)
:For(J,1,C)
:[B](I,1)*[C](1,J)/T→[E](I,J)
:End
:End
:For(I,1,R)
:[B](I,1)→[E](I,C+1)
:[B](I,1)→[A](I,C+1)
:End
:For(J,1,C)
:[C](1,J)→[E](R+1,J)
:[C](1,J)→[A](R+1,J)
:End
:T→[E](R+1,C+1)
:T→[A](R+1,C+1)
:For(I,1,R)
:For(J,1,C)
:([A](I,J)-
[E](I,J))² /[E](I,J)→[D](I,J)
:End
:End
:0→S
:For(I,1,R)
:For(J,1,C)
:S+[D](I,J)→S
:End
:End
:ClrHome
:Disp "TEST STAT = ",round(S,4)
:Disp "D OF F = ",(R-1)*(C-1)
:S→T
:(R-1)*(C-1)→D
:Disp "COMPUTE P-VALUE?"
:Disp "1=YES, 2=NO"
:Input J
:If J=2:Stop
:ClrHome
:Disp "TEST STAT = ",round(T,4)
:Disp "D OF F",D
:prgmZPVALUE
:Disp "EXPECTED VALUES?"
:Input "1=YES,2=NO",M
:If M=2
:Then:Goto N
:End
:ClrHome
:For(I,1,R)
:Disp "ROW",I
:For(J,1,C)
:Disp [E](I,J)
:End
:Disp "PRESS ENTER"
:Pause
:ClrHome
:End
:Lbl N
:Stop
```

Program **GOODFT82.82P**

```
:ClrHome
:ClrList L4,L5,L6
:Disp "GOODNESS-OF-FIT"
:Disp "TEST"
:Disp " "
:Disp "ENTER NUMBER"
:Input "OF OUTCOMES ",N
:ClrHome
:Disp "ENTER "
:Disp "OBSERVED VALUES"
:For(I,1,N)
:Input X
:X→L4(I)
:End
:ClrHome
:Disp "ENTER "
:Disp "EXPECTED VALUES"
:For(I,1,N)
:Input X
:X→L5(I)
:End
:For(I,1,N)
:(L4(I)-L5(I))²/L5(I)→L6(I)
:End
:ClrHome
:Disp "TEST STAT = "
:sum L6→S
:Disp round(S,4)
:Disp "DEG OF F=",N-1
:Disp "COMPUTE P-VALUE?"
:Disp "1=YES, 2=NO"
:Input J
:If J=2:Stop
:ClrHome
:S→T
:N-1→D
:Disp "TEST STAT = ",round(T,4)
:Disp "DEG OF F=",D
:prgmZPVALUE
:Stop
```

Program **HYPTESTB.82P**

```
:ClrHome
:Disp "FINDS TEST STAT."
:Disp "AND P-VALUE FOR"
:Disp "B, SLOPE OF "
:Disp "REGRESSION LINE"
:Disp " "
:Disp "IND=L1,DEP=L2"
:Disp "CONTINUE?
:Input "1=YES   2=NO",S
:If S=2:Stop
:dim L1→N
:sum (L1*L2)-sum L1*sum L2/N→A
:sum L1²-(sum L1)²/N→B
:sum L2²-(sum L2)²/N→C
:A/B→F
:√((C-F*A)/(N-2))→E
:E/√B→G
:F/G→T
:N-2→D
:ClrHome
:Disp "TEST STAT =",T
:Disp "D OF F = ",D
:Disp "COMPUTE P-VALUE?"
:Input "1=YES 2=NO",U
:If U=2:Stop
:ClrHome
:Disp "TEST STAT =",T
:Disp "D OF F = ",D
:prgmZPVALUE
```

Program **ONEPOPCI.82P**

```
:ClrHome:Disp "FIND THE C.I."
:Disp "ONE POPULATION"
:Disp " "
:Disp "1 = MEAN"
:Disp "2 = PROPORTION"
:Input "CHOICE ",C
:ClrHome
:If C=1
:Then
:Menu("SOURCE OF
DATA","L1",1,"STATS",2,"STOP
NOW",Ø)
:Lbl Ø:Stop
:Lbl 1
:mean(L1)→X
:dim (L1)→N
:√(sum (L1-X)² /(N-1))→S
:Disp "x̄ = ",X
:Disp "S = ",S
:Disp "n = ",N
:Disp "PRESS ENTER"
:Pause
:ClrHome
:Goto 3
:Lbl 2
:Input "x̄ = ",X
:Input "S = ",S
:Input "n = ",N
:Lbl 3
:Disp "1 = LARGE SAMPLE"
:Disp "2 = SMALL SAMPLE"
:Input "CHOICE ",D
:If D=1
:Then
:ClrHome
:Disp "ENTER Z"
:Input "FOR CONF.INT",Z
:Goto A
:End
:ClrHome
:Disp "ENTER T"
:Input "FOR CONF.INT",Z
:Lbl A
:ClrHome
:Disp "LOWER BOUND:"
:X-Z*S/√N→L
:Disp round(L,4)
:Disp "UPPER BOUND: "
:X+Z*S/√N→U
:Disp round(U,4)
:Goto B
:End
:Input "P-HAT = ",P
:Input "n = ",N
:1-P→Q
:ClrHome
:Disp "ENTER Z"
:Input "FOR CONF.INT",Z
:ClrHome
:Disp "LOWER BOUND:"
:P-Z*√(P*Q/N)→L
:Disp round(L,3)
::Disp "UPPER BOUND: "
:P+Z*√(P*Q/N)→U
:Disp round(U,3)
:Lbl B
:Stop
```

Program **ONEPOPHT.82P**

```
:ClrHome
:Disp "FINDS TEST STAT"
:Disp "ONE POPULATION"
:Disp " "
:Disp "1 = MEAN"
:Disp "2 = PROPORTION"
:Input "CHOICE",C
:ClrHome
:If C=1
:Then
:Menu("SOURCE OF
DATA","L1",1,"STATS",2,"STOP
NOW",Ø)
:Lbl Ø:Stop
:Lbl 1
:mean(L1)→X
:dim (L1)→N
:√(sum (L1-X)²/(N-1))→S
:Input "HO MEAN = ",M
:Goto B
:Lbl 2
:Input "HO MEAN = ",M
:Input "S(OR SIGMA) = ",S
:Input "x̄ = ",X
:Input "n = ",N
:Lbl B
:ClrHome
:Disp "x̄ =",X
:Disp "ST.DEV. =",S
:Disp "ST DEV OF x̄ = ",S/√N
:Disp "PRESS ENTER "
:Pause
:ClrHome
:(X-M)/(S/√N)→T
:Disp "TEST STAT = ",round(T,4)
:Goto A
:End
:Input "HO PROPORTION = ",H
:Input "P-HAT = ",P
:Input "n = ",N
:1-H→Q
:ClrHome
::Disp "ST DEV OF P-HAT =
",√(H*Q/N)
:(P-H)/√(H*Q/N)→T
:Disp "TEST STAT = ",round(T,4)
:Lbl A
:Disp "COMPUTE P-VALUE?"
:Disp "1=YES, 2=NO"
:Input J
:If J=2:Stop
:ClrHome
:Disp "TEST STAT = ",round(T,4)
:N-1→D
:prgmZPVALUE
:Stop
```

Program **POISSN82.82P**

```
:ClrHome
:Menu("POISSON PROBABILITY","SINGLE
P(X)",1,"SUM OF MANY",2)
:Lbl 1
:Input "ENTER LAMBDA ",L
:Disp ""
:Input "     ENTER X ",X
:(L^X*e^-L)/(X!)→A
:Disp "     P(X) = ",round(A,4)
:Stop
:Lbl 2
:Ø→C
:Input "ENTER LAMBDA ",L
:Input "ENTER MIN X  ",A
:Input "ENTER MAX X  ",B
:For(X,A,B)
:(L^X*e^-L)/(X!)→D
:D+C→C
:End
:Disp "PROBABILITY =",round(C,4)
:Stop
```

Program **PROBDIST.82P**

```
:ClrHome
:ClrList L3,L4,L5,L6
:Disp "FINDS MEAN AND"
:Disp "STANDARD DEVIATION"
:Disp "OF A PROBABILITY"
:Disp "DISTRIBUTION"
:Disp " "
:Input "NO.OF OUTCOMES ?",N
:ClrHome
:Disp "ENTER X VALUES"
:For(I,1,N)
:Input X
:X→L3(I)
:End
:ClrHome
:Disp "ENTER P(X) VALUES"
:For(I,1,N)
:Input P
:P→L4(I)
:End
:ClrHome
:For(I,1,N)
:L3(I)*L4(I)→L5(I)
:(L3(I))²*L4(I)→L6(I)
:End
:sum L5→M
:sum L6→P
:Disp "MEAN = ",M
:Disp " "
:Disp "ST DEV = ",√(P-M²)
:Stop
```

Program **RANDOMNO.82P**

```
:ClrHome
:Disp "GENERATES RANDOM"
:Disp "POS. INTEGERS"
:Disp " "
:Disp "ENTER N,"
:Disp "THE LARGEST NO."
:Input N
:ClrHome
:Disp "TO STOP PROGRAM"
:Disp "PRESS 'ON'"
:Disp "TO CONTINUE"
:Disp "PRESS ENTER"
:Lbl A
:1+int (rand*N)→H
:Disp H
:Pause
:Goto A
```

Program **TWOPOP.82P**

```
:ClrHome
:Disp "TWO POPULATIONS "
:Disp "    FIND"
:Disp "1=CONF. INT."
:Disp "2=TEST STAT "
:Disp "3=BOTH"
:Input E
:ClrHome
:Disp "1 =  MEANS"
:Disp "2 =  PROPORTIONS"
:Input "CHOICE ",C
:If C=2
:Then
:Goto I
:End
:ClrHome
:Menu("SOURCE OF
DATA:","POP1=L1 POP2=L2",1,"STATS",2
,"STOP NOW",Ø)
:Lbl Ø:Stop
:Lbl 1
:mean(L1)→M
:dim (L1)→N
:√(sum (L1-M)² /(N-1))→S
:mean(L2)→U
:dim (L2)→W
:√(sum (L2-U)² /(W-1))→V
:Disp "x̄1 =",M
:Disp "S1 =",S
:Disp "n1 =",N
:Disp "PRESS ENTER"
:Pause
::Disp "x̄2 =",U
:Disp "S2 =",V
:Disp "n2 =",W
:Disp "PRESS ENTER"
:Pause
:ClrHome
:Goto C
:Lbl 2
:Input "x̄1 = ",M
:Input "S1 = ",S
:Input "n1 = ",N
:Input "x̄2 = ",U
:Input "S2 = ",V
:Input "n2 = ",W
:ClrHome
:Lbl C
:Disp "1= Z-SAMPLE"
:Disp "2= T-SAMPLE"
:Input "CHOICE ",J
:If J=1
:Then
:ClrHome
:√(S² /N+V² /W)→X
:If E=2:Goto 4
```

```
:Disp "ENTER Z "
:Input "FOR CONF INT",Z
:Goto A
:End
:Disp "1 = SIGMAS ="
:Disp "2 = SIGMAS ≠"
:Input "CHOICE ",F
:If F=1
:Then:Goto B
:End
:ClrHome
:Disp "D OF F ="
:int ((S² /N+V² /W)² /((S² /N)² /(N-
1)+(V² /W)² /(W-1))→D
:Disp D
:√(S² /N+V² /W)→X
:If E=2:Goto 4
:Disp "ENTER T"
:Input "FOR CONF INT",Z
:Lbl H
:M-U-Z*X→L
:M-U+Z*X→Y
:Goto G
:Lbl A
:√(S² /N+V² /W)→X
:M-U-Z*X→L
:M-U+Z*X→Y
:ClrHome
:Lbl G
:Disp "CONF. INTERVAL"
:Disp "MEAN 1 - MEAN2"
:Disp "LOWER BOUND = ":Disp
round(L,4)
:Disp "UPPER BOUND = ":Disp
round(Y,4)
:Disp "PRESS ENTER"
:Pause
:ClrHome
:If E=1:Stop
:ClrHome
:Lbl 4
:Disp "TEST STAT ="
:(M-U)/X→T
:Disp round(T,4)
:Goto K
:Stop
:Lbl B
:ClrHome
:√(((N-1)*S² +(W-1)*V² )/(N+W-
2))*√(1/N+1/W)→X
:Disp "D OF F = "
:N+W-2→D
:Disp D
:If E=2:Goto 4
:Disp "ENTER T"
:Input "FOR CONF. INT ",Z
```

(CONTINUED ON NEXT PAGE)

Program **TWOPOP.82P**

(CONTINUED FROM PREVIOUS PAGE)

```
:Goto H
:End
:Lbl I
:ClrHome
:Input "P-HAT 1 = ",P
:Input "n1 = ",R
:Input "P-HAT 2 = ",S
:Input "n2 = ",Y
:If E=2:Goto M
:Disp "ENTERZ"
:Input "FOR CONF INT =  ",Z
:ClrHome
:P-S-Z*√((P*(1-P))/R+(S*(1-S))/Y)→L
:P-S+Z*√((P*(1-P))/R+(S*(1-S))/Y)→U
:Disp "CONF. INTERVAL"
:Disp "P1 - P2 "
:Disp " "
:Disp "LOWER BOUND = ":Disp
round(L,4)
:Disp "UPPER BOUND = ":Disp
round(U,4)
:Disp " "
:Disp "PRESS ENTER"
:Pause
:ClrHome
:If E=1:Stop
:Lbl M
:ClrHome
:√((R*P+Y*S)/(R+Y)*(1-
(R*P+Y*S)/(R+Y))*(1/R+1/Y))→X
:Disp "TEST STAT = "
:(P-S)/X→T
:Disp round(T,4)
:Lbl K
:Disp "COMPUTE P-VALUE?"
:Disp "1=YES, 2=NO"
:Input J
:If J=2:Stop
:ClrHome
:Disp "TEST STAT = ",round(T,4)
:prgmZPVALUE
:Stop
```

Program **ZCHIAREA.82P**

```
:D/2→A
:prgmZGAMMA
:A*G→G
:"X^A*e^-X/G→Y1
:Y1(T/2)+fnInt(Y1,X,Ø,T/2)
:1-Ans
:Disp "P-VALUE=",round(Ans,4)
```

Program **ZGAMMA.82P**

```
:If fPart A=Ø
:Then
:(A-1)!→G
:Else
:√π→G
:.5→X
:While X<A-.5
:XG→G
:X+1→X
:End
:End
```

Program **ZPVALUE.82P**

```
:Menu("TYPE OF
DISTRIBUTION","NORMAL",1,"T-
",2,"CHI-SQ",3)
:Lbl 1:"(√(2π))^-1*e^(-.5*X²)"→Y1
:Menu("TYPE OF TEST","RIGHT-
TAILED",4,"LEFT-TAILED",5,"TWO-
TAILED",6)
:Lbl 4
:fnInt(Y1,X,T,5)→P
:Goto A
:Lbl 5
:fnInt(Y1,X,⁻5,T)→P
:Goto A
:Lbl 6
:fnInt(Y1,X,abs (T),5)*2→P
:Lbl A
:Disp "P-VALUE = ",round(P,4)
:Return
:Lbl 2
:prgmZTAREA
:Return
:Lbl 3
:prgmZCHIAREA
:Return
```

Program **ZTAREA.82P**

```
:D/2→A
:prgmZGAMMA
:G→C:(D+1)/2² a
:prgmZGAMMA
:⁻A→A
:fnInt((1+X²/D)^A,X,Ø,abs (T))
:Menu("TYPE OF TEST","RIGHT-
TAILED",1,"LEFT-TAILED",2,"TWO-
TAILED",3)
:Lbl 1
:If T<Ø
:Then
:.5+AnsG/C√Dπ
:Else
:.5-AnsG/C√Dπ
:End
:Disp "P-VALUE = ",round(Ans,4)
:Goto 5
:Lbl 2
:If T>Ø
:Then
:.5+AnsG/C√Dπ
:Else
:.5-AnsG/C√Dπ
:End
:Disp "P-VALUE = ",round(Ans,4)
:Goto 5
:Lbl 3
:.5-AnsG/C√Dπ
:Disp "P-VALUE = ",round(2*Ans,4)
:Lbl 5
```

Index

Notes

Notes

Notes

Notes

Notes

Notes

Notes

Notes

Notes

Notes

Notes

Notes

Notes

Notes